WORKSHOP PHYSICS® ACTIVITY GUIDE

Activity-Based Learning

MODULE 3: HEAT, TEMPERATURE, AND NUCLEAR RADIATION

Thermodynamics, Kinetic Theory,
Heat Engines, Nuclear Decay, and Radon Monitoring
(Units 16-18 & 28)

PRISCILLA W. LAWS
DICKINSON COLLEGE

with contributing authors:
ROBERT J. BOYLE
PATRICK J. COONEY
KENNETH L. LAWS
JOHN W. LUETZELSCHWAB
DAVID R. SOKOLOFF
RONALD K. THORNTON

WILEY

JOHN WILEY & SONS, INC.

Cover Image: James Fraher/Image Bank/Getty Images

To order books or for customer service, please call 1-800-CALL-WILEY (225-5945).

ISBN 0-471-64163-4

Printed in the United States of America

SKY10030097_092421

CONTENTS

CONTENTS

UNIT 16: TEMPERATURE AND HEAT TRANSFER

The coffee and the bath water are at the same temperature. Both are "hotter" than the air in the room. However, it costs much less to heat the coffee than it does to heat the bath water. Why? Aren't the words heat and temperature synonymous? Is heat a substance or what? In this unit you will learn how contemporary physicists define and use familiar terms like "heat" and "temperature" to help them understand thermal processes.

UNIT 16: TEMPERATURE AND HEAT TRANSFER

. . . . thermometer readings alone do not tell the entire story of thermal interaction, . . . something else must be happening, and . . . an additional concept . . . must be invented.

Arnold Arons
A Guide to Introductory Physics Teaching
Ch. 5, p. 120 (1990)

OBJECTIVES

1. To acquire an operational definition of temperature and to understand the connection between temperature and thermal equilibrium.

2. To understand how the temperature of a body changes as it undergoes thermal interaction with its surroundings.

3. To quantify the interaction between parts of a system that leads to temperature changes as a function of the mass and of the system.

Fig. 16.1.

16.1. OVERVIEW

With this unit we begin the study of *thermodynamics*, a new and profoundly different way of studying physical phenomena. Much of the physics studied in previous units involved motions that we could see, while many of the changes we will encounter in thermodynamics will not be visible without the help of indirect measuring instruments such as thermometers and manometers. We will use some of the concepts from mechanics, such as work and kinetic energy, in our discussion of thermodynamics, but we will introduce some new terms as well. Although you have already encountered some of the terms used in the science of thermodynamics, such as *heat transfer* and *temperature*, we will eventually define them more precisely. Other new terms such as *entropy*, *adiabatic process*, and *isothermal process* that will be introduced in the next few units are less familiar.

Temperature is one of the most familiar and fundamental thermodynamic quantities and it is the major focus of study in this first unit on thermodynamics. In general, the measurement of temperature depends on some characteristic of material in a thermometer changing as it is heated or cooled. Thus, the length of a metal rod, the height of a column of mercury, or the volume of a gas under pressure can serve as means of measuring temperature. When we study electrical phenomena, you will also discover that the electrical current carried by certain materials can change with temperature. Thus, it is possible to use electronic devices to measure temperature. In the early sections of this unit you will use both the familiar glass bulb thermometer and an electronic temperature sensor interfaced to a computer data acquisition system to measure temperature.

In later activities you will take a much more careful look at the concept of temperature. In particular you will observe how the temperature of a substance or system is affected when it interacts with an environment or another substance at a different temperature. Whenever the temperature of something changes, we say glibly that it has undergone a *thermal interaction*. When the temperature of a system remains constant, we refer to it as being in *thermal equilibrium*. Since we cannot see what really goes on when something changes temperature, we have to develop some new concepts to try to explain what is happening. One of these new concepts is that of *thermal energy transfer* (also commonly known as heat transfer).

In the last sections you will combine ideas about temperature change, thermal energy transfer, and internal energy to develop the First Law of Thermodynamics. An understanding of this law is essential to the practical application of thermodynamics to heat engines and other phenomena of vital economic importance. It is also another piece of the energy conservation puzzle that is crucial to our understanding of all physical phenomena.

MEASURING TEMPERATURES

16.2. MEASURING TEMPERATURE

Various thermometers have been devised that allow for reliable and quantitative temperature measurements. The most common type of laboratory thermometer consists of a glass tube filled with colored alcohol or mercury, a metal that is a liquid at ordinary temperatures. To begin our study of thermodynamics we will crudely define temperature as *a quantity that is related to the height of a liquid inside a familiar glass bulb thermometer.*

Fig. 16.2.

At your lab station you will find a thermometer without a scale. Fiddle around with it, using substances of apparently different temperatures around the room, and see if you can figure out how it works. To conduct this investigation and several others in this section you will need:

- 1 unmarked glass thermometer
- 1 masking tape (to mark temperature scales)
- 1 glass thermometer with F and C scales (–5°C to 105°C)
- 1 Hot Pot (for boiling water)
- 4 containers, 0.5 liter (for crushed ice, salt, hot tap water, and cold tap water)
- crushed ice
- salt (NaCl)
- 2 Styrofoam cups
- 2 Styrofoam cup lids
- 1 vat (to prevent spilling)

Recommended group size:	2	Interactive demo OK?:	N

16.2.1. Activity: The Glass Bulb Thermometer

Sketch a picture of a glass bulb thermometer. Explain how it works.

16.3. TEMPERATURE SCALES

You are probably familiar with the Fahrenheit scale from weather forecasts and you may have worked with the Celsius scale in other science courses. These are just two of the many temperature scales set up by early investigators of heat and thermodynamics. Most of these scales were set up by taking two "fixed points" of reliable, easily reproducible temperatures. The investigator then had to decide how many "degrees" lay between these two temperatures.

The Fahrenheit scale was developed by a German physicist, Gabriel Fahrenheit in 1724. The zero point (0°F) of his scale was supposed to be the lowest temperature attainable with a mixture of ice and salt, while the upper point was human body temperature that he called 96°F. On his original scale the freezing point of water exposed to air at sea level turned out to be about 32°F (the ice point) and the boiling point of water exposed to air at sea level turned out to be about 212°F (the steam point).

In 1742 a Swedish investigator named Celsius devised another scale that he referred to as the centigrade scale. On this scale the freezing point of water exposed to air at sea level was fixed at 0°C and the boiling point of water exposed to air at sea level was fixed at 100°C. Today we use a modification of the original scale that is based on a degree that is the same "size" as the centigrade scale, but is fixed so that the *triple point* of water is 0.01°C. The triple point is defined as the temperature and pressure at which solid ice, water, and vapor can coexist.

Physicists have discovered that there is a natural limit to how cold any object can get. This coldest possible temperature, now called absolute zero, is the zero point of another temperature scale, the Kelvin scale. Since it is based on a "true" zero point, this is a very important temperature scale for thermodynamics. This scale has the same "size" degree as the Celsius scale, and on this scale the triple point of water is 273.16K (exactly). Conversion between the Celsius scale and the Kelvin scale is very simple since one merely needs to add 273.15 to the temperature in degrees C to get the Kelvin temperature. In other words,

$$T_K = T_C + 273.15$$

Conversely,

$$T_C = T_K - 273.15$$

16.3.1. Activity: Defining a Temperature Scale Operationally

a. Define a temperature scale of your own or use one of the standard scales. To do this you should pick *two* objects whose temperatures you suspect are different and seem convenient to measure, assign different values, known as *fixed points*, to the temperatures of these two objects, and then decide how many "degrees" should lie between these assigned values. Use a strip of masking tape on the thermometer to mark your scale. Describe how you set up your scale.

b. On your scale, determine the room temperature, ice water temperature, and your body temperature (arm pit). Also determine these values using the marked thermometer for comparison.

	Your temperature scale	Degrees Celsius
Room		
Ice water		
Body		

c. Are the fixed points you chose reliable? In other words, are they truly *fixed* points? Could you, if given more time and better apparatus, have chosen more reliable ones? Explain. Give an example of a more reliable fixed point.

16.3.2. Activity: Converting Between Temperature Scales

a. You should derive an equation that converts a temperature expressed on the Fahrenheit scale to a temperature expressed on the Celsius scale. Start by writing down two different values of temperature expressed in both scales of two fixed points (e.g., the ice and steam points, although you could use any two points whose temperatures you know on both scales).

Point #1:_____ : _____°F = _____°C

Point #2:_____ : _____°F = _____°C

Now assume that our equation is a simple linear relation of the form:

$$Y = mX + b \text{ or } T_C = mT_F + b$$

[Celsius] = (slope)[Fahrenheit] + constant

1. The slope m tells us how many Celsius degrees there are for each Fahrenheit degree. You can find out what m is by comparing $T(\text{steam}) - T(\text{ice})$ (or other fixed point temperatures) in both scales.

$$m = \underline{\hspace{2cm}} \,°C/°F$$

2. The constant b can be solved for by noting that when the Celsius scale is 0°C (the ice point) the Fahrenheit scale is 32°F.

$$b = \underline{\hspace{2cm}} \,°C$$

3. The final relation is:

b. Using the same procedure, write down an equation that converts a temperature expressed on the scale you developed in Activity 16.3.1 to the Celsius scale.

c. Express the temperature of this room on the *Kelvin* scale.

16.4. SENSING TEMPERATURE ELECTRONICALLY

Next you will explore temperature measurement with an electronic temperature sensor that can be attached directly to a computer data aquisition system. This system has several advantages over the use of the glass bulb thermometers. The sensors usually respond more quickly to changes in temperature. You can produce a graph of temperature vs. time for one or two sensors at a time automatically. And, as usual, the data you collect can be displayed in tabular form and analyzed. Or you can transfer your data to other programs for further analysis and display. The purpose of this activity is to become familiar with electronic temperature measurement, some limitations of electronic sensing, and features of the data acquisition software you'll need to use in future activities.

To carry out these investigations you will need:

- 1 computer data acquisition system
- 1 temperature sensor
- 1 glass thermometer with F and C scales (-5° C to 105° C)
- 2 containers, 0.5 liter (for ice water and hot tap water)
- crushed ice

Recommended group size:	3	Interactive demo OK?:	N

To get started:

1. *Open the data acquisition software* file L160401 or set up your own file to display a digital readout of temperature.
2. *Display the temperature of the air around you* by starting the temperature vs. time data collection.

Electronic vs. Glass Bulb Temperatures

Typically electronic sensors are set (calibrated) so they yield readings that are similar to those given by glass bulb thermometers. If not, you may need to calibrate your sensor before starting the next activity. If needed, you can learn how to calibrate your electronic temperature sensor by using the software help feature.

16.4.1. Activity: Comparing Temperatures

a. To check the accuracy of your electronic sensor, you can remeasure a couple of temperatures using both the electronic sensor and glass bulb thermometer. Try ice water and warm tap water and fill out the table below.

	Glass bulb therm (°C)	Electronic therm (°C)	Difference (°C)
Ice water			
Warm water			

b. How close is the electronic reading to your glass bulb thermometer reading? **Note:** If you are off by more than ±1 degree Celsius, you should calibrate your sensor carefully whenever accurate temperature readings are needed.

Some Important Properties of Temperature Sensing

There are a couple of things you should know about temperature sensing in order to measure temperature more accurately.

16.4.2. Activity: Time Delays

a. When someone pops a thermometer which is at room temperature in your mouth to see if you have a fever, can your temperature be determined immediately? Why not?

Fig. 16.3.

b. Suppose you want to measure room temperature with a thermometer that has been in ice water. Which do you predict would cause more time delay—measuring room temperature water or room temperature air? Explain the reason for your prediction.

c. Use the graphing feature of your temperature software to verify your prediction *quantitatively*. To do this, record how the temperature of an electronic temperature sensor changes over time when it is transferred from ice water to room air and then to room temperature water. Then determine the time it takes for the sensor to reach room temperature in each case.

Ice water to room air: Δt (sec) =

Ice water to room temp water: Δt (sec) =

d. On the basis of these measurements what should you watch out for in making temperature measurements?

e. The temperature difference between room temperature and ice water is about 20°C. What do you think will happen to the measured time delays if the temperature of the sensor is only a degree or two below room temperature? **Hint:** Your temperature vs. time graphs contain the answer.

Remember how it feels to get out of a shower on a cool dry day? Brrr! You need to beware of cooling by evaporation. Be careful not to measure air temperatures when the sensitive part of the thermometer is wet (especially with alcohol). Evaporating liquid on the thermometer can cool it. You will be in a better position to understand and explain the phenomenon of cooling by evaporation after completing the next couple of units on thermodynamics in which we study the relationship between temperature and molecular motion in a substance.

16.5. THERMAL EQUILIBRIUM

Are objects lying around a room really at the same temperature? To explore this question of thermal equilibrium you can use several objects which have holes drilled in them for a thermometer bulb.

- 1 piece of metal with hole
- 1 piece of Styrofoam with hole
- 1 piece of wood with hole
- 1 glass thermometer

Recommended group size:	2	Interactive demo OK?:	N

16.5.1. Activity: Predicting Relative Temperatures

a. Feel the wood, metal, and Styrofoam. Predict which object actually has the highest temperature and the lowest temperature.

b. Now measure the temperature of the three objects and record your measurements in the table below.

Metal	(°C)
Wood	(°C)
Styrofoam	(°C)

c. Did your observation jibe with your prediction? Is your sense of touch an accurate predictor of relative temperatures?

d. According to other observations you have made in this activity, should the temperatures near the surface of three different materials sitting around in the same room be the same or different?

e. On the basis of previous observations, you should be able to explain the reason why some objects feel colder than others. **Hint:** Is the temperature of your hand different from the room temperature? If so, what is happening when you touch a room temperature object?

TEMPERATURE CHANGES AND INTERACTIONS

16.6. DOES TEMPERATURE TELL THE WHOLE STORY?

We know that when a hotter substance comes into thermal contact with a cooler one the temperatures of the two substances change. These temperature changes are easy to observe when the substances can either mix or come into thermal contact with each other. Do the initial temperatures alone allow us to predict the final temperature of the system after the two substances have interacted with each other?

Suppose you have two liquids of masses m_A and m_B in thermal contact inside a fairly well-insulated container. You will be asked to make some predictions and test them with some measurements. For this activity you will need:

Fig. 16.4. Should you add cream to coffee right away?

- 1 computer data acquisition system
- 1 electronic temperature sensor
- 1 Styrofoam cup
- 1 Styrofoam cup lid
- 1 thin-walled aluminum tube (approx. 1.5"outer dia. × 3.25" high)
- 1 rubber stopper (no holes, for tube bottom)
- 1 rubber stopper (one hole, for tube lid)
- 2 containers, 0.5 liter (for hot and cool water)
- 1 container (for crushed ice)
- crushed ice
- 1 electronic scale (or 1 graduated cylinder, 200 ml)

Note: For water, 1 ml = 1 cc = 1 g.

Recommended group size:	4	Interactive demo OK?:	N

---

*This can be constructed using industrial aluminum tubing with 0.035" wall thickness. Rubber stoppers are used to seal the tube ends.

16.6.1. Activity: Predicting Temperature Changes

a. If you were to place two equal masses of the same type of liquid having different temperatures in thermal contact, how would you determine the final temperature? For example, suppose $m_A = m_B = 32$ g while $T_A = 5°C$ and $T_B = 35°C$. What would you expect to happen to the temperature of each of the liquids after a while due to thermal contact between them? **Note:** Assume that the liquids are inside an insulated container so no interaction takes place with the room.

b. If you were to place two *different* masses of the same type of liquid having different temperatures in thermal contact, how would you determine the final temperature? For example suppose $m_A = 50$ g and $m_B = 200$ g while $T_A = 5°C$ and $T_B = 30°C$. What would you expect to happen to the temperature of each of the liquids after a while due to thermal contact between them? **Note:** If you can't make a quantitative prediction, try making a *qualitative* prediction. For example, will the liquids in both vessels undergo the same temperature *change*?

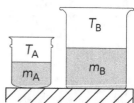

Fig. 16.5. Two vessels containing different masses of water at different temperatures.

c. Explain the reasons for your answers in parts a. and b. On the basis of what you already know, what do you think is taking place when the two liquids come into thermal contact? Can you describe a possible mechanism for any interactions you might predict? Do you think it's possible on the basis of knowing just the temperatures, *but not the masses*, of the two liquids in the containers to predict the final temperatures of the two liquids after they have been in contact?

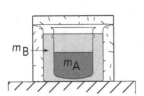

Fig. 16.6. Two vessels of water in thermal contact surrounded by an insulator such as a covered Styrofoam cup.

d. Use the equipment listed above or other available apparatus to test your prediction in part b. This can be done by filling the aluminum tube with a mass, m_A, of hot water. This tube can then be placed in an insulated container with very cold water of mass m_B. This is shown in Figure 16.6. Electronic temperature sensors should be placed in each container of water. Monitoring the temperatures on a real time graph for about 3 minutes should be sufficient. Briefly describe what you did and your results. If possible, you should share your results with others. **Note:** To compensate for the lack of perfect insulation of the system from its surroundings, use about 4 or 5 times as much chilled water in the Styrofoam cup as you have hot water in the aluminum tube.

e. How did your observations agree with your prediction? Is there any evidence of the liquids mixing together or exchanging matter during the thermal interaction? Is there any visible exchange of matter? Explain any new ideas about how the thermal interaction might be causing the temperatures to change.

Parts of any insulated system can be in thermal contact with each other without mixing. If these parts have different temperatures, they will interact until the entire system is at the same temperature. This is a mysterious process, because the *interaction that causes temperature changes in two parts of a system can occur without an exchange of matter.*

You should have noticed from your experiments and those of your classmates that the relative masses of the parts of your thermally isolated system affect the value of the final equilibrium temperature. Thus, the interaction between two parts of a system cannot be explained as a simple temperature change. We need to create a new concept to help us understand heating and cooling processes. Scientists have invented the concept of thermal energy transfer (or heat transfer) to explain this phenomenon.

The use of the noun "heat" is misleading, since using this term to explain temperature changes implies the exchange of a substance between two parts of a system. The word "heat" is actually a sloppy shorthand for an interaction process that often leads to temperature changes. As a reminder that we are dealing with a process rather than a substance, we are going to refer to thermal energy transfer (or sometimes heat transfer) and not simply *heat* for the remainder of this unit.

What is thermal energy transfer? Can it occur without the exchange of matter? In the remainder of this unit we will endeavor to understand more about the nature of thermal energy transfer, to explore the possibility that it is a form of thermal energy exchange, and to quantify the amount of thermal energy or heat transfer occurring in different processes. In the next unit you will explore a model for explaining thermal energy exchange (or so-called heat transfer processes) on an atomic level.

16.7. COOLING RATES

Before exploring quantitative aspects of thermal energy exchange (heat transfer) and its nature in the next sections, let's study the rate at which heating and cooling take place under different conditions.

We all know that a cup of hot water will eventually cool down in a room while a cup of ice water will warm up. What is the final temperature of warm water in a room? What does the rate at which the temperature of an object changes depend on? Expressed mathematically, we are asking the question: What does the transfer rate, R, given by the derivative of the temperature, T, with respect to time, t, depend on?

$$R = \frac{dT}{dt} \qquad (16.1)$$

Let's do a thought experiment! Imagine that you transfer thermal energy to any kind of liquid (water, syrup, mercury, antifreeze, and so on) and place it in any type of container. Suppose that you have a large, well-insulated room that can be maintained at any reasonable temperature. Thus, you could chill the room to 0°C (brrr) or warm it up to 35°C (phew!).

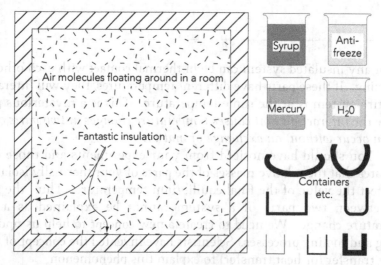

Fig. 16.7. Types of containers and liquids that might be used in experiments to measure cooling rates in an insulated room.

16.7.1. Activity: Predicting Relative Cooling Rates

a. If a small amount of a substance that is at 50°C is placed in a large insulated room, what will the final temperature of the substance be? Will heat transfer to the room take place? Will the room temperature go up? Explain your answer on the basis of your predictions and observations in Activity 16.6.1 parts b. and c.

b. List at least four or five variables that the rate of cooling of an object in a large room might depend on.

c. Describe a situation in which you expect the initial cooling rate of a given object to be rapid and one in which the initial cooling rate might be slow. **Note:** The *initial cooling rate* is the change in temperature per second at first and not the total time it takes something to cool.

Does Container Material Effect Cooling Rate?

One variable that the cooling rate of a liquid might depend on is the material that the walls of the container are made of. In order to explore the effect of the container, you can use the following equipment:

- 1 computer data acquisition system
- 1 temperature sensor (calibrated)
- 1 thin-walled aluminum tube (approx. 1.5" outer dia. × 3.25" high)
- 1 plastic bottle, 2 oz. (approx. 1.5" outer dia. × 3.25" high)
- 1 rubber stopper (no holes, for tube bottom)
- 1 rubber stopper (one hole, for tube lid)
- 1 rubber stopper (one hole, for bottle lid)
- 2 Styrofoam cups
- 1 vat (to prevent spilling)
- crushed ice

Recommended group size:	4	Interactive demo OK?:	N

16.7.2. Activity: Exploring Cooling Rates

a. Suppose you placed equal masses of hot tap water into the plastic bottle and aluminum tube. If each container was immersed in a mixture of ice and water, which container would allow the water to cool faster?

b. Set your temperature sensor to monitor the cooling of each vessel of water for about 4 minutes (or use the experiment file L160702). Set your data rate to about 1 data point per 5 seconds or you will have an overwhelming amount of data. You can immerse each can in a large cup filled with ice and water. *Be sure to shake or stir the liquid in its container during the entire cooling period.* Affix a printout of an overlay graph of your two cooling curves in the space that follows. **Note:** Transfer your temperature data to a spreadsheet and *save it* as you may need it for homework.

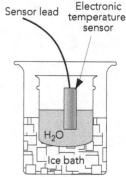

Fig. 16.8. Setup for measuring the cooling rate of water electronically.

c. Was your prediction correct? Explain.

Note: In the situation you just studied, the time rate of decrease in temperature of an object is usually proportional to the temperature difference between the object and its surroundings at each time as it cools. This is known as *Newton's Law of Cooling*. It can be shown mathematically that the dependence of the temperature, T, of an object as a function of elapsed time, t, is given by:

$$T(t) = (T_i - T_s)e^{-\alpha t} + T_s$$

where T_s is the temperature of the surroundings,

$\quad\quad T_i$ is the initial temperature of the object,

and $\quad \alpha$ is the cooling constant which depends on a number of factors for a given system. It has dimensions of inverse time.

THERMAL TRANSFER AS AN ENERGY EXCHANGE

16.8. IS THERMAL ENERGY TRANSFER REALLY AN ENERGY TRANSFER?

So far you have made observations that indicate that interactions take place when two substances in thermal contact are at different temperatures. We have dubbed these interactions "thermal energy transfer." In this Activity Guide and in most textbooks, you are told glibly that thermal energy (or heat) transfer is an energy transfer. What evidence do you have that this is so?

16.8.1. Activity: Speculations on Thermal Energy Transfer as Energy Exchange

Discuss any reasons that you and your classmates can think of for identifying thermal energy transfer with energy transfer (other than "the text says so"!). This is not an obvious identification because we have no way of seeing what is going on.

Let's try an observation that might help make the identification of thermal energy transfer with energy plausible. This observation involves the use of a hand-operated device to heat your finger. For this observation you will need the following:

- 1 mini-generator (hand-operated)
- 2 alligator clip leads (for the generator)
- 1 miniature light bulb
- 1 light bulb socket
- 1 nichrome wire, approximately 18" long

Recommended group size:	2	Interactive demo OK?:	N

You should wrap the wire loosely around your index finger, attach it to the generator, and have your partner turn the crank.

16.8.2. Activity: Doing Mechanical Work to "Heat" Your Finger

a. What happens to your wire-wrapped finger when the mini-generator crank is being turned?

Fig. 16.9.

b. Remember the definition of work? Are you exerting a force on the handle as you turn the crank? If so, is the handle moving in the direction of the force you are exerting? Are you doing mechanical work when you turn the crank? Why or why not?

c. Is it less work to turn the crank when the wire is not attached to the generator? What happens when a light bulb is attached?

d. What is probably happening to the energy you expend doing the work to turn the crank? What types of energy transformations might be taking place when the finger is wrapped with wire?

Clearly, when the bulb lights as you crank the generator, mechanical energy is causing the generation of electrical energy. Although you might not have studied electrical energy yet in a formal way, from what you observed above you can see that the electrical energy produced by the mechanical work resulted in a thermal transfer that raises your finger's temperature. Thus, it seems appropriate to associate the heating of your finger with energy exchange. In the next set of activities you will be using electrical energy as a mechanism for the quantitative study of thermal energy transfer in relation to temperature change for different types of substances.

The idea of heating as an energy transfer process that results from temperature differences leads to more formal definitions of thermal energy exchange and of temperature as a measure of thermal equilibrium. These are summarized below. Although we ask you to memorize very few things in this course, these definitions should be memorized and their meanings understood. Retaining these two important thermodynamic concepts is critical to mastering the next unit on the First Law of Thermodynamics, the ideal gas law, and heat engines.

---

1. Thermal energy transfer or exchange is energy in transit between two systems in thermal contact due only to temperature difference with the hotter system losing energy as the cooler system gains it.
2. Two objects are in thermal equilibrium, and hence have the same temperature, if no net energy is exchanged between them when they are placed in thermal contact.

16.9.   THE HEAT PULSER–A NEW RESEARCH INSTRUMENT

The computer-based temperature-sensing system can be outfitted with an electrical heat pulser. This pulser can be activated by hitting a key on the computer keyboard. Hitting the "pulse" button on the computer screen will turn on a pulse for a time period that you can set in the software. The power in watts (the rate at which the pulser delivers energy) is determined by the power rating of the pulser unit. Since the power, P, is energy per unit time, you can calculate the amount of thermal energy transfer, ΔQ, in joules delivered to a substance each time you activate the heat pulser. Recall that 1 watt = 1 joule/second.

If Δt is the preset time for a heat pulse, then

$$P = \frac{\Delta Q}{\Delta t} \quad \text{or} \quad \Delta Q = P\Delta t$$

The heat pulser is extremely valuable in quantifying the relationship between thermal energy transfer and temperature change for different substances. Let's experiment with it a bit. To do this you will need:

- 1 computer data acquisition system
- 1 temperature sensor
- 1 heat pulser
- 1 immersion heater, ~ 200 W
- 1 Styrofoam cup (and water)
- 1 vat (to prevent spilling)

Recommended group size:	4	Interactive demo OK?:	N

Warning: Never turn on a heat pulser unless the immersion heater coil is completely submerged in water. Otherwise it will burn out!

16.9.1. Activity: The Pulser, Heating, and Energy Output

a. Open up the data acquisition software and use the experiment file L160901—or set up your software for temperature measurements. If necessary, calibrate your temperature sensor. Next, submerse the temperature sensor and immersion coil in water in an insulated Styrofoam cup. What happens to the temperature of the water when you hit the on-screen pulse button to get a 10 s pulse from the immersion heater?

b. Earlier you observed the cooling rates of the hot water in an aluminum can. How might the heat pulser be used to keep the temperature of the hot water from decreasing?

c. What is the time that electricity flows in the immersion heater for each pulse of "heat"?

$$\Delta t = \underline{\hspace{2cm}} \text{ s}$$

d. What is the power rating in watts listed on the heat pulser unit?

$$P = \quad\quad\quad \text{watts}$$

e. Assuming that all the electrical energy contributed by the heat pulser results in thermal energy transfer to the liquid it is immersed in, calculate the thermal energy in joules transferred by one pulse. **Note:** Some of the electrical energy is transferred to the immersion coil also.

f. How much thermal energy will be transferred if the pulse time is doubled each time a pulse is activated?

16.10. THERMAL ENERGY TRANSFER, TEMPERATURE, AND SPECIFIC HEAT

If you transfer pulses of energy thermally to a perfectly insulated cup of some liquid, what determines how much temperature change, ΔT, takes place? How does ΔT depend on:

1. The number of pulses of thermal energy you transfer? [ΔQ^{total}]
2. The mass of liquid in the cup? [m]
3. The kind of liquid you have?

You should conduct a series of observations in which you demonstrate *quantitatively* that if the cup is *well insulated*:

$$\Delta Q^{total} = c \, m \, \Delta T$$

where c is a constant that depends on the kind of liquid you have. To do this project you will need to investigate the three factors by changing only one variable at a time. For example, you can use the same mass of room temperature water for a series of experiments and vary only the amount of heat you add. Then you can use the same amount of heat and vary the mass of the water. Finally, you can use the same mass of liquid and the same amount of heat and vary the type of liquid (that is, use antifreeze as one liquid and water as the other). You should determine the values of c for both water (c_w) and antifreeze (c_a).

To do the series of observations you will have the following equipment available to you:

- 1 computer data acquisition system
- 1 temperature sensor
- 1 heat pulser
- 1 immersion heater, 200 W
- 2 Styrofoam cups
- 2 plastic cup lids
- 2 ceramic coffee mugs or 2 calorimetry cups
  (to keep the Styrofoam cups from tipping)
- 2 containers, 0.5 liter (for water at room temperature)
- antifreeze, 250 ml
- 1 electronic balance
- 1 vat (to prevent spilling)

Recommended group size:	4	Interactive demo OK?:	N

By using the real time graphical display of temperature changes in L161001 or setting up your own experiment file, you can study the relationship between ΔQ and ΔT in one run as long as the temperature pulses are added far enough apart to allow each resulting temperature change to be recorded before the next pulse is added.

> **Warnings:** Use enough liquid in each case to make sure the electric coil is covered in every observation. Be careful not to use large amounts of liquid since the heating process will take too long! Keep stirring the liquid at all times. *Do not heat the antifreeze over 70°C at any time during the experiment!*

16.10.1. Activity: The Relationship of Thermal Energy Transfer and Temperature

a. List data and affix any relevant graphs in the space that follows. Describe your observations and findings. Please be as quantitative as possible.

b. What substance changes temperature the most for a given amount of thermal energy transfer—the water or the antifreeze? Which one has the higher value of c? Be sure to list appropriate units for c_a and c_w.

Back in the good old days when a kid was given a hot baked potato to carry to school on a cold winter day, the potato kept her warm and served as lunch! Why is a hot potato a better kid warmer that a small bag of popcorn? If a body made of a given substance has a certain thermal energy transfer, how much will its temperature increase? The answers to these questions involve a property of a substance called the *specific heat*. The constant c that you determined for water and antifreeze is known as the specific heat of a substance. *Specific heat is defined in J/kg · °C units as the amount of thermal energy transfer in joules needed to raise a kilogram of a substance by one degree Celsius.*

Fig. 16.10.

The mathematical definition of specific heat is constructed, in fact, by turning Equation 16.2 around to get

$$c \equiv \frac{Q}{m\Delta T} \qquad (16.3)$$

A material with a high specific heat can have a large quantity of thermal energy transferred to it without changing its temperature very much. A material of the same mass with a small heat capacity will undergo the same temperature change when a smaller amount of heat is transferred to it.

UNIT 17: THE FIRST LAW OF THERMODYNAMICS

Keith Bell/Dreamstime.com

Objects can only start rising in the atmosphere if there is an upward force on them that is greater that the gravitational force pulling them downward. They must be lighter than air. Each of these balloons has a burner under it to heat air that fills it. Why is hot air at atmospheric pressure lighter than cold air? How will the mass of the air in a full balloon vary with temperature? How much heat energy must be transferred to a given balloon before it will lift off? In this unit you will learn how to describe the behavior of a gas mathematically when energy is transferred to it and why it becomes lighter.

UNIT 17: THE FIRST LAW OF THERMODYNAMICS

It must be admitted, I think, that the laws of thermodynamics have a different feel than most of the other laws of the physicist. There is something more palpably verbal about them—they smell more of their human origin.

P. W. Bridgman (1941)

OBJECTIVES

1. To observe conditions under which thermal energy transfer can cause a substance to change phase but not temperature.

2. To explore how thermal energy transferred to a gas can cause the gas to expand and do work on its surroundings.

3. To learn about how hidden *internal energy* in a substance is related to atomic and molecular motions and to the potential energy they can store.

4. To learn how the First Law of Thermodynamics can be used to relate the internal energy of a substance to thermal energy exchange and the work done if the substance expands or contracts.

5. To discover the relationship between the pressure, volume and temperature in "ideal" gases that cannot store internal potential energy.

6. To develop a simple molecular model to explain gas behavior.

17.1. OVERVIEW

In the last unit you explored situations in which two substances in thermal contact changed their temperatures without exchanging matter. We attributed these temperature changes to an exchange of thermal energy. But what is this thermal energy exchange that causes a substance of heat up or cool down? In this unit we will develop the concept that all substances store hidden internal energy in their atoms or molecules. We will use this concept of internal energy to explain how thermal energy transfer works and other phenomena that you have not yet explored.

In Section 17.3 you will observe that when H_2O is changing phase (from ice to water or from water to steam) you can transfer thermal energy to a substance without changing its temperature. We will then use the concept of internal energy to explain both why temperatures don't change in phase transitions and why they do change in other circumstances.

In Section 17.4 you will investigate how thermal energy transfer can cause air, which is a gas, to expand and do mechanical work on its surroundings by raising a piston. This is a situation in which the energy transferred to the system is not all stored internally. Some of it is transformed into work.

Then in Section 17.6 you will use the principle of conservation of energy to formulate the first law of thermodynamics. A belief in this law allows us to calculate how much the internal energy stored in a system is increased when work is done on it *or* thermal energy is transferred to it. An understanding of the first law of thermodynamics is important in practical endeavors such as the design of heat engines. It is also used in fundamental scientific research.

In Sections 17.7 through 17.9 you will investigate what happens to air at about room temperature when heat energy is transferred to it and when when it is compressed or expanded. Although we could never hope to follow the motion of each and every particle in a gas, you will learn about certain macroscopic properties of a gas that can be easily measured in our laboratory. These *macroscopic* properties are temperature, pressure, and volume and the mathematical relationship between these properties that you will discover is known as the ideal gas law.

Finally, in Sections 17.10 through 17.12, you will use Newton's laws to explain how macroscopic quantities might be connected to the forces, velocities, and momenta of the billions upon billions of particles we believe are contained in gases whose internal energy is only accounted for by the kinetic energy of its atoms and molecules. Thus, you will construct an idealized *microscopic* picture of a gas and explain the ideal gas laws in terms of the motion of the gas molecules.

INTERNAL ENERGY, WORK, AND THERMAL ENERGY

17.2. THERMAL ENERGY TRANSFER WITHOUT TEMPERATURE CHANGE

As part of our quest to understand more about thermal energy transfer and the internal energy of a substance, let's look at the question of whether energy transfer can take place without a temperature change occurring. Just in case you haven't memorized this yet, let's return to the principle of heat energy transfer we developed in the last unit:

> Thermal energy exchange takes place between two systems in thermal contact when there is a temperature difference between them.

Heat, or caloric as it was called in the old days, used to be thought of as a substance. Even today the term heat is often used casually in a way that implies that it is a substance. Even though scientists now use the term heat as a shorthand term for a thermal energy transfer process rather than a substance, we will continue to use the term thermal energy transfer to remind you that the transfer of energy from one system to another does not necessarily involve the exchange of matter.

Let's return to the question at hand. Is it possible for a system to absorb thermal energy without changing temperature?

17.2.1. Activity: Thermal Energy Absorption Without Temperature Change?

a. From your experiences with the heating and cooling of different substances, can you think of any situations in which a system has been in thermal contact with something at a higher temperature and not changed its temperature?

b. For the examples you have picked, can you think of any internal changes going on in the system that could help explain the lack of temperature rise?

17.3. CHANGING ICE TO WATER AND THEN TO STEAM

Melting ice and boiling water were probably examples that came up in your discussions about substances that can absorb thermal energy without changing temperature. Let's make some predictions about melting and boiling.

17.3.1. Activity: Predicting *T* vs. *t* for Water

a. Suppose you added energy thermally at a constant rate to a container of water at 0°C (with no ice in it) for 20 minutes at a low enough rate that the water almost reaches its boiling point. Sketch the predicted shape of the heating "curve" on the following graph.

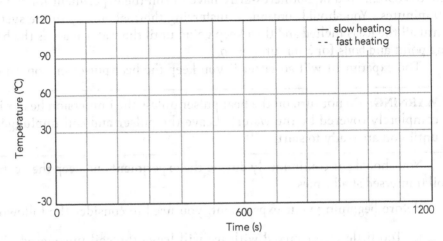

b. Suppose that the container had a mixture of ice and water at 0°C when you started heating it at a faster rate and that the water starts boiling after five minutes (300 seconds). You keep adding energy at the same rate for five more minutes. Draw a dotted line on the preceding graph showing your prediction.

Determining Heats of Fusion and Vaporization

Let's observe what actually happens to temperatures when thermal energy is transferred to a mixture of ice and water at a continuous rate. By transferring a known amount of energy to a mixture that is originally half water and half ice, we can also determine the amount of thermal energy needed to melt a gram of ice. This energy is known as the *latent heat of fusion*. You can also determine the amount of energy needed to turn a gram of boiling water into steam. This energy is known as the *latent heat of vaporization*. It is often measured in joules per gram. For this activity you will need:

- 1 computer data acquisition system
- 1 temperature sensor (calibrated)
- 1 heat pulser
- 1 immersion heater, approximately 200 W
- 1 Styrofoam cup, 300 ml
- 1 electronic balance (or 200 ml graduated cylinder)
- 2 containers, approximately 250 ml (for crushed ice and water at 0°C)
- crushed ice
- paper towels (to dry the ice)

Recommended group size:	2	Interactive demo OK?:	N

> **Note:** If needed, you should calibrate the temperature sensor using hot and cold water before you begin this experiment.

Set up the temperature sensing software to take temperature vs. time data for 20 minutes while the heat pulser is enabled (or use the experiment file L170302). You do not necessarily have to run the experiment for the full 20 minutes. You should continue transferring thermal energy to the system until all the ice is melted and then keep going until the water reaches the boiling point and boils for 5 minutes or so.

The experiment will go faster if you keep the heat pulser on constantly.

> **WARNING!** Do not turn on the heat pulser unless the immersion heater is completely covered by the water! Leave the pulser and coil unplugged until you are ready to start.

You should stir continuously during the experiment and keep the heating coil immersed at all times.

Before beginning your experiment, you need to consider the following:

1. Too little water mixed with ice will leave the coil uncovered. Too much water/ice mixture will take a long time to heat up.
2. How can you remove as much water as possible from the crushed ice before mixing it with the ice water? (Paper towels help.)
3. How can you chill the water so it is 0°C before the dry crushed ice is mixed with it?
4. How can you create a mixture of ice water and crushed ice with roughly equal masses of each?
5. In order to calculate the latent heats of fusion and vaporization you must determine the rate at which your immersion heater transfers thermal energy to water. How can you use the specific heat of water which is 4190 J/kg·C and the time it takes the melted ice and water to boil to calculate the J/s delivered by the heater?
6. Once you determine the immersion heater output in J/s,
   (a) how can you calculate how many joules of energy you added to the ice and water mixture during the time the ice was melting?
   (b) How can you measure how many grams of water changed to steam during the time the water was boiling?

Note: After your group has devised a plan for doing the experiment and figured out what measurements you will need to take, *you should review your plans with your instructor before starting to take data.*

17.3.2. Activity: The Actual *T* vs. *t* for Water

a. What was the initial mass of the ice in grams? What will the final mass of the water be after all the ice has melted?

b. In the following diagram, sketch or affix the graph that resulted from transferring thermal energy (in other words, "heating") at a constant rate to the approximately 50/50 ice and water mixture for about 15 or 20 minutes (until the original water–ice mixture heated to boiling and then boiled for about five minutes).

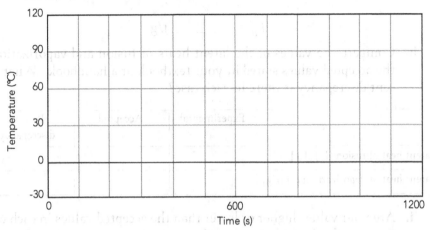

c. Analyze your graph from the time the ice has melted until the water just starts to boil. Use your knowledge of the mass of the melted ice and water mixture along with the known value for the specific heat of water to calculate how many joules per second are delivered by the immersion heater.

d. How long did it take the ice to melt? Use your result from part c. to calculate how much energy in joules was added by the heat pulser while the ice was melting.

e. Remembering how many grams of ice you started with, calculate L_f, the latent heat of fusion of water. In other words, how many joules per gram are needed to melt ice? Show your steps.

$$L_f = \underline{\hspace{1.5cm}} \text{J/g}$$

f. How many minutes did you allow the water to boil before you stopped the experiment? How many grams of water turned to steam in that time? How much heat energy was transferred to the water in that time?

g. What is the value of the Latent Heat of Vaporization? Show your calculations!

$$L_v = \underline{\hspace{2cm}} \text{ J/g}$$

h. Compare the values of the latent heats of fusion and vaporization to the accepted values stated in your textbook or a handbook. What percent discrepancy is there in each case?

	Experimental	Accepted	Percent discrepancy
Latent heat of fusion (L_f) [J/g]			
Latent heat of vaporization(L_v) [J/g]			

i. Are your values higher or lower than the accepted values in each case? Can you think of any sources of systematic error to explain this?

Internal Energy

If you lift a mass or compress a spring, it is obvious that a conservative system gains an amount of potential energy, usually denoted ΔU. Is the concept of "potential energy" useful in discussing what happens if work is done on a system such as gas confined in a syringe where there is no apparent change in potential energy? The answer is yes, but we have to give a new meaning to our potential energy. In thermodynamics, it is called a change in the *internal energy* of a system, ΔE^{int}. Internal energy is defined as the invisible microscopic energy stored in a system. In complex systems, this microscopic energy can consist of both kinetic energy (due to the translational, vibrational, and rotational motions of atoms and molecules) and potential energy (such as that stored in chemical bonds).

States of Matter

Fig. 17.1.

Most substances can exist in three states—solid, liquid, and gas. Some gases, such as helium, become liquid only at extremely low temperatures and some solids, such as a diamond, are very hard to melt. Usually, these *changes of state* or *phase changes* require a transfer of thermal energy. During a phase change, the substance can absorb thermal energy (or transfer thermal energy to its surroundings) without changing the temperature of the substance until the phase change is complete. The standard explanation for this is that the thermal energy is transferred to a system is used to raise its potential energy and breaks the interatomic or intermolecular bonds that characterize a phase. But once a phase change has taken place, additional thermal energy is used to increase the kinetic energy of atoms and molecules in the system if we associate an increase in the system's hidden kinetic energy with its temperature.

17.4. WORK DONE BY AN EXPANDING GAS

One system we will meet often in our study of thermodynamics is a mass of gas confined in a cylinder with a movable plunger or piston. The use of a gas-filled cylinder to study thermodynamics is not surprising since the development of thermodynamics in the eighteenth and nineteenth centuries was closely tied to the development of the steam engine, which employed hot steam confined in just such a cylinder.

You are to observe the relationship between expansion and compression of a gas and work done on or by the gas. To do this you will need the following:

- 1 glass syringe, 10 cc
- 1 length Tygon® tubing, 5 cm (1/8" ID)
- 1 tubing clamp

Recommended group size:	2	Interactive demo OK?:	N

Raise the syringe plunger about halfway up and insert the short tube and clamp at the end of the syringe to seal it.

Try compressing the air in the syringe gently. Do you have to do work on the gas to compress it? What happens when the plunger springs back?

In thermodynamics, pressure (defined as the component of force that is perpendicular to a given surface per unit area of that surface) is often a more useful quantity than force alone. It can be represented by the equation:

$$P = \frac{F_\perp}{A}$$

Let's extend our definition of work developed earlier in the course and this new definition of pressure to see if we can calculate the work done by a gas on its surroundings as it expands out against the piston with a (possibly changing) pressure P.*

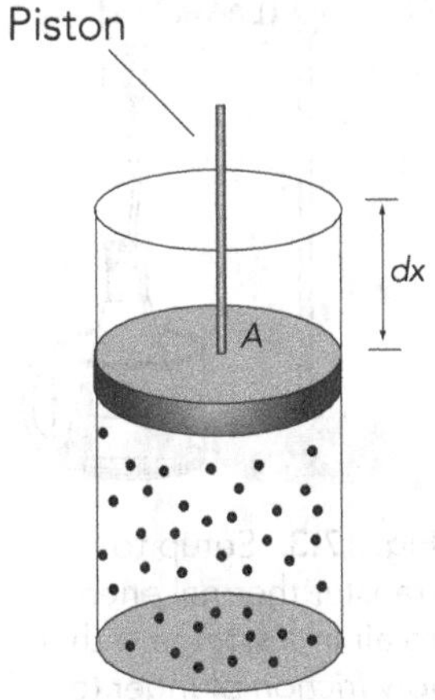

17.4.1. Activity: Relating Work and Pressure Mathematically

a. Are you doing work when you compress the gas in a syringe?

b. You know that work can be written as $W = \int F dx$ when F represents the force you exert on the syringe plunger. Show the mathematical steps to verify that work can be written as $W = \int P dV$ for the situation shown in Figure 17.1.

Piston

Fig. 17.2. Gas at pressure P exerts a force on the piston of $F = PA$ as it moves a distance dx.

17.5. WORK AND THERMAL ENERGY TRANSFER

As you observed in the last activity, you can do work and compress a gas. But then where did the gas get the energy to do work on the piston? It must have come from an increase in the internal energy you gave it when you did work on it. So, when the compressed gas is allowed to expand, it can do physical

*We use capital P to represent pressure to distinguish it from momentum which is represented by small p.

work on its surroundings by raising a piston. One way to increase the internal energy of a system is to do work on it. Suppose that instead of doing work on a system, you transfer thermal energy to it?

Thermal Energy Transfer and Work on Surroundings

Let's consider a system consisting of just the air inside a syringe, tube, and flask that are connected to each other. Transferring thermal energy to the system could serve to increase its internal energy. Alternatively, it could cause a system to do work on its surroundings and leave its internal energy unchanged. In thermodynamics we are interested in the relationship between heat energy transfer to a system's internal energy and work done by the system on its surroundings. What do you think would happen if you attach a cylinder with a low-friction movable piston (or plunger) to a small flask and place the flask in hot water? Would its piston experience a force? Can the air in the system do work on the plunger? For this activity you will need:

- 1 glass syringe, 10 cc*   • 1 #5 one-hole rubber stopper
- 1 test tube clamp   • 1 length Tygon® tubing, 30 cm (1/8" ID)
- 1 rod   • 1 coffee mug (for hot tap water)
- 1 rod stand (or table clamp) • 1 tray (to prevent spilling)
- 1 Erlenmeyer flask, 125 ml

Recommended group size:	2	Interactive demo OK?:	N

17.5.1. Activity: The Heated Syringe

a. Predict what happens to a plunger (piston) if a flask attached to it is put into hot tap water that is about 40-50° C as shown in Fig. 17.3.

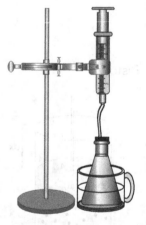

b. Now try it! Clamp a syringe and attach it to the flask with its plunger partway down. *While holding the plunger fixed,* submerge the flask in hot tap water. Explain what might be happening to the gas while you hold the plunger to keep the gas volume constant? Could its internal energy be changing? Explain.

Fig. 17.3. Setup to transfer thermal energy to air in a syringe with a *low friction* plunger (or piston) in it.

c. After a minute or so, release the plunger and let it move freely. What happens? Is this what you predicted would happen in part a. above?

d. Is the system doing work on its surroundings? Why or why not?

---

* This can be done as a demonstration using the PASCO scientific Heat Engine/Gas Law Apparatus (TD-8572).

17.6. THE FIRST LAW OF THERMODYNAMICS

What is the relationship between heat energy transfer, changes in the system's internal energy, and the work done by (or on) the system? We denote E^{int} as the total hidden internal energy in the system. To help us understand how E^{int} is related to work and thermal energy, we will consider the gas confined in a syringe and flask in Activity 17.5.1 to be our system and the flask of water as its surroundings. For simplicity, we will neglect the flask and syringe and assume that thermal energy transfer is just between the system's air and the water surrounding the flask.

Suppose the plunger on the syringe is clamped while the flask attached to it is immersed in hot water. The clamped plunger can't move, so no work is done since the water is at a higher temperature than the system air, thermal energy is transferred from the water to the air. This causes the temperature of the water to decrease and the temperature of the air trapped in the syringe and flask to increase. If energy is conserved, the thermal energy transferred to the air can be calculated using the equation $Q = c_w m \Delta T$ where m_w is the mass of the water, c_w is its specific heat, and ΔT is the temperature change of the water.

Assume that no thermal energy can be transferred to the surroundings. Then if no work is done on or by the system, the transferred thermal energy, Q, must equal the *increase* in the internal energy of the system air. This assumption is based on a *belief* that energy is conserved in the interaction between the hot water and the trapped air.

Suppose we release the plunger and allow the air to expand and do work on the plunger when it's placed in the hot water. How can we calculate the work done by the air and its change in internal energy?

As the trapped air expands we can calculate the amount of work it did on its surroundings by evaluating the integral you derived in Activity 17.4.1b, $W = \int P dV$. Where did the energy to do this work come from? The only possible source is the internal energy of the air, which must have decreased by an amount W. The total change in the internal energy of our trapped air must be

$$\Delta E^{int} = Q - W \qquad (W = \text{work } by \text{ system}) \qquad (17.1)$$

This relationship between absorbed thermal energy, work done on surroundings, and internal energy change is believed to hold for any system, not just for air trapped in a syringe and a flask. It is known as the *first law of thermodynamics*.

The first law of thermodynamics has been developed by physicists based on a set of very powerful inferences about energy forms and their transformations. We ask you to accept it on faith. The concepts of work, thermal energy transfer, and internal energy are subtle and complex. For example, work is not simply the motion of the center of mass of a rigid object or the movement of a person in the context of the first law. Instead, we have to learn to draw system boundaries and total the mechanical work done by the system inside a boundary on its external surroundings.

The first law of thermodynamics is a very general statement of conservation of energy for thermal systems. It is not easy to verify it in an introductory physics laboratory, and it is not derivable from Newton's laws. Instead, it is an independent assertion about the nature of the physical world based on a belief that energy in the universe is conserved.

More Comments About the First Law

There are many ways to achieve an internal energy change, ΔE^{int}. To achieve a small change in the internal energy of air in a syringe, you could transfer a large amount of thermal energy to it and then allow the gas to do work on its surroundings. Alternatively, you could transfer a small amount of thermal energy to the air and not let it do any work at all. The change in internal energy, ΔE^{int}, could be the same. Since ΔE^{int} depends only on $Q - W$ and not Q or W alone, it is said to be "path independent."

17.6.1. Activity: The First Law

a. Write down *in words only* your understanding of the First Law of Thermodynamics.

b. Can you think of any situations where W is negligible and $\Delta E^{int} = Q$? (**Hint**: Is it necessary to extract work from a cup of hot coffee to cool it? Can you think of similar situations?)

c. How could you arrange a situation where Q is negligible and in which $\Delta E^{int} = -W$? Such situations have a special name in thermodynamics. They are called *adiabatic processes*.

THE IDEAL GAS LAW

17.7. CHARACTERISTICS OF GASES

Thermal Behavior of an Ideal Gas

Any gas can be described by the macroscopic variables volume V, pressure P, and temperature T. Today's activities will help us understand how P, V, and T are related for an ideal gas. In particular we will carry out an experiment to relate V and T, another to relate P and T, and a third to relate P and V. We will then combine the relationships we discover into the *ideal gas law*. In a later section we will study how this law might be connected to a picture of an ideal gas as a collection of particles moving about in a container.

Measuring Gas Pressure with a Manometer (Optional)

Note: You should complete this section if you do not have an electronic pressure sensor and a computer data acquisition system.

Before undertaking the study of the ideal gas law, we need to learn something about how pressure is measured. At your lab station there should be a "manometer" that looks like the one pictured in Figure 17.4. The air close to the surface of the earth exerts a force on each unit of area it encounters. This is atmospheric pressure, which is often denoted P^{atm}. The columns of water on each side of the manometer experience this atmospheric pressure. If one column is higher than the other, there is an additional pressure at the bottom of the column due the weight of the liquid that is above the level of the water in the lower column.

For this activity you will need to construct a manometer as shown in Figure 17.4 using:

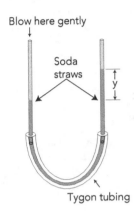

Blow here gently

Soda straws

Tygon tubing

Fig. 17.4. A simple manometer for measuring pressures.

- 2 translucent soda straws, 8 cm (1/4" OD)
- 1 length of Tygon® tubing, 10 cm (1/4" ID)

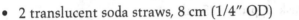

Recommended group size:	2	Interactive demo OK?:	N

17.7.1. Activity: Pressure of a Column of Liquid

a. Put some water in the bottom of the manometer tube. Try blowing into one end, first very lightly and then somewhat more strongly. (It is a good idea for each partner to blow into a different straw to keep from spreading germs.) What changes? How might you use this change to measure pressure?

b. Pressure is defined as the force per unit area; that is, $P = F/A$. Show that the *additional* pressure, ΔP, exerted by a column of liquid of height y is independent of the cross-sectional area, A, of the column and is given by $\Delta P = \rho g y$. **Hints:** Recall that the density of a liquid is given by mass per unit volume; that is, $\rho = m/V$. Start by finding the extra force in the diagram due to the weight of the column of water. Ignore the air pressure above the column of water in making this calculation.

17.8. DETERMINING THE IDEAL GAS LAW

It turns out that between about 0°C and 50°C, air behaves approximately like an ideal gas. Each group will study one of several possible relationships between P, V, and T by investigating: how V changes with T (Charles' law I); how P changes with T (Charles' law II); or how P changes with V (Boyle's law). By sharing the results of these three investigations, you should be able to determine a simple equation that relates P, V, and T. See if you can "discover" Boyle's or Charles' laws by doing project A, B, or C described below. You should share data and findings with your classmates and fill in the results of all three experiments in Activities 17.8.1 through 17.8.3.

PROJECT A: V vs. T for a Gas (Charles' Law I)

How does temperature affect the volume occupied by a trapped gas maintained at a constant pressure? For this activity you will need:

- 1 glass syringe, 10 cc* (w/ Luer lock tip)
- 1 Erlenmeyer flask, approx. 125 ml
- a computer-data acquisition system
- 1 # 5 one-hole rubber stopper
- 1 electronic temperature sensor or a thermometer, 0–50°C
- 1 length of Tygon® tubing, approx. 30 cm (1/8" ID)
- 1 beaker (500 ml)
- 1 immersion heater (200 W)
- 1 rod stand
- 1 rod
- 1 test tube clamp
- 1 small tray (to prevent spilling)

Recommended group size:	3	Interactive demo OK?:	N

You should set up the apparatus shown in Figure 17.5. and put enough cold water in the beaker to cover the immersion heater with cold water. To prevent spills, put the beaker on the tray. Place the tubing in the rubber stopper hole. Place the thermometer in the mug outside the flask. Since you will be heating the water with the immersion heater to increase the volume of the trapped air, push the plunger in to about the 1 cc mark before seating the rubber stopper in the flask. Mount the syringe horizontally as shown in the diagram.

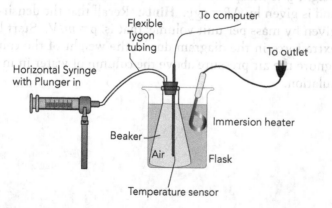

Fig. 17.5. Diagram of apparatus that can be used to measure volume changes as a function of temperature in air trapped at a constant pressure.

---

* A good supplier for this syringe is Aldrich, catalog # Z10,106-0. Project A can be also demonstrated to the class or done by an individual group using the PASCO scientific Heat Engine/Gas Law Apparatus (TD-8572).

> **Note:** Since the hose has a cylindrical shape, you can calculate the volume of air contained in it by noting that the equation for the volume of a cylindrical shape of length L and inner diameter d is given by $V = \pi(d/2)^2 L$.   (17.2)

17.8.1. Activity: Volume vs. Temperature

a. How do you predict that the volume of the air trapped in the flask, hose, and syringe will change if it is heated (or cooled) at constant pressure? Sketch a graph and explain your prediction.

b. Explain why the pressure of the air remains constant even if the volume of the air in the system changes.

c. You will start measurements of V vs. T with the plunger bottomed at the 0 or 1cc mark when the flask is sitting in the cold water. Determine the initial volume of the system including the hose, flask, and syringe. Show your basic measurements and calculations. **Hint:** You can use an electric balance to determine how many grams of water it takes to fill the flask up to the stopper. The number of grams is the same as the number of cubic centimeters of available flask volume.

Volume of hose: _____ cm^3

Volume of flask: _____ cm^3

d. Stir the water vigorously as you take and record the data needed to determine the total volume of the trapped air as a function of the temperature of the water surrounding the gas inside the flask. **Note:** Strictly speaking, you should measure the temperature of the air in the flask, but it's much faster to measure the surrounding water temperature which is roughly proportional to that of the air.

T(C)	T(K)	V flask (cm^3)	V syringe (cm^3)	V total (cm^3)
			10	

e. Use the Kelvin temperature scale and affix a graph of your V vs. T data (or the graph from another group). How are V and T related mathematically?

PROJECT B: *P* vs. *T* for a Gas (Charles Law II)

What happens to the pressure, P, of a gas if its volume is kept constant while its temperature, T, changes? For this project you will need:

- 1 computer data acquisition system
- 1 data logger software (C170802)
- 1 immersion heater
- 1 Erlenmeyer flask, approx. 125 ml
- 1 #5 one-hole rubber stopper
- 1 plastic in-line coupler (for connecting tubing to pressure sensor)
- 1 lengths of Tygon® tubing, 50 cm (1/8" inner diameter)
- 1 beaker (500 ml)
- 1 temperature sensor (or thermometer, 0-50°C)
- 1 pressure sensor (to measure small pressure changes)

Recommended group size:	3	Interactive demo OK?:	N

We want you to measure how gas pressure depends on its temperature. In other words, you should determine the shape of a P vs. T curve at constant volume and display your results using a computer graphing routine.

The flask contains your test volume of a gas (air). You can begin the experiment by immersing the flask in the beaker filled with cold water. Then, as you transfer thermal energy to the water with the immersion heater you can measure the pressure of the air in the flask as a function of its temperature.

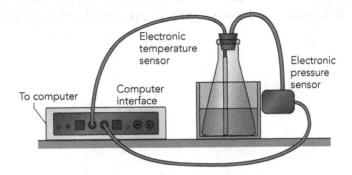

Fig. 17.6. Apparatus to measure pressure vs. temperature.

Notes: (1) Configure the data logger software using "Event Mode."
(2) The pressure sensor data should have 15 point averaging to smooth
out fluctuations. (3) Since we are only trying to determine if a linear relationship between P and T exists, it is probably ok to:
  a. Not bother calibrating the sensors.
  b. Measure the temperature of the water surrounding the flask in-
     stead of the air inside the flask. The two temperatures should be
     roughly proportional to each other, and measuring water temper-
     ature takes much less time.

Use the immersion heater to warm the water in the beaker slowly and
measure the pressure until the temperature of the surrounding water stabi-
lizes. **Note:** You should stir the water vigorously while taking data.

17.8.2. Activity: *P* vs. *T*

a. Sketch what you think a graph of P vs. T might look like.

b. Enter your data (*or summarize data from other groups*), explain your
   calculations, and show your graphical display of P in atm or N/m^2 vs.
   T in degrees Celsius below.

c. How does P depend on T if V is held constant? Use the graph to find the best equation. How does this compare with your prediction?

PROJECT C: Boyle's Law (P vs. V)

How does the pressure of a gas depend on its volume at a constant temperature? A syringe connected to an electronic pressure sensor can be used to measure how a change in pressure affects the volume of a gas at room temperature. To verify Boyle's law you will need:

- 1 plastic disposable syringe, 10 cc
- 1 computer data acquisition system
- 1 pressure sensor (up to 2 atm)

Recommended group size:	2	Interactive demo OK?:	N

The approach to obtaining measurements is to trap a volume of air in the syringe and then compress the air slowly to smaller and smaller volumes by pushing in the plunger. The gas should be compressed slowly so it will always have time to come into thermal equilibrium with the room (and thus be at room temperature). You should take pressure data for about 10 different volumes.

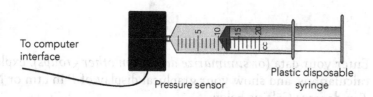

To computer interface

Pressure sensor

Plastic disposable syringe

Fig. 17.7. Plastic disposable syringe and pressure sensor used to verify the relationship between pressure and volume for a gas.

If possible, attach the end of the syringe directly to the electronic pressure sensor. Set up the sensor, interface, computer, and software to record pressures. Use the experiment file L170803 or set up an event mode to record pressure automatically and volume manually. Place the plunger at about the halfway mark on the syringe and then attach it to the pressure sensor to seal the system. Next, pull the plunger out to the 10 cc mark and start recording data at 10 cc, 9 cc, and so on. You can determine each volume in cc visually and enter its value into the computer.

---

*If this is not possible, use short sections of hose and take the volume of air in it into account in calculations.

17.8.3. Activity: P vs. V

a. Sketch your prediction for the shape of a graph of pressure vs. volume for the air in the syringe and connecting tubing.

b. Enter your data (*or summarize data from other groups*), explain your calculations, and show your graphical display of P in atm or N/m^2 vs. V in cm^3.

c. Try modeling or fitting your P, V relation by predicting a relationship between P and V such as P vs. V^2 or P vs. $1/V$ and so on and affix the best fit below.

d. How did the shape of your graph compare with your prediction?

17.9. THE IDEAL GAS LAW

We have seen how pressure depends on temperature at constant volume, and how volume depends on temperature at constant pressure, and how pressure and volume are related at a constant temperature. Let's summarize these relationships using simple mathematics.

17.9.1. Activity: Summarizing Boyle's and Charles' Laws

a. Write down an equation that describes the relationship (as explored in Activity 17.8.1) between volume and temperature when the pressure is held constant. Express this in terms of V, T, and C_1, where C_1 is a constant. Then solve the equation for C_1.

b. Write down an equation that describes the Charles' law II relationship you discovered in Activity 17.8.2 in terms of P, T, and another constant, C_2. Then solve the equation for C_2.

c. Write down an equation that describes the Boyle's Law relationship you discovered in Activity 17.8.3 in terms of P, V, and another constant, C_3. Then solve the equation for C_3.

d. Multiply the three relationships from parts a., b., and c. together to get the product $C_1 C_2 C_3$ on one side and a combination of P, V, and T on the other side.

e. Now take the square root of both sides and summarize the results in terms of a new constant C_4. Solve your equation so the product PV is on the left and C_4 and T are on the right.

The Ideal Gas Law describes all three relationships mathematically in an idealized fashion. If you did the same experiments with different amounts of gas, you would find that the factor $\sqrt{C_4}$ turns out to be nR. So the Ideal Gas Law is given by:

$$PV = nRT \qquad\qquad (17.3)$$

where n = the number of moles of gas
           R = the Universal Gas Constant given by 8.31 J/mol·K

An alternative statement of the Ideal Gas Law is:

$$PV = Nk_BT \qquad\qquad (17.4)$$

where N = the number of gas molecules
           k_B = Boltzmann's Constant given by 1.38×10^{-23} J/K

17.9.2. Activity: The Ideal Gas Law

a. Show that all three relationships observed in the Activities in Section 17.8. are compatible with the Ideal Gas Law.

b. Describe the Ideal Gas Law in words.

KINETIC THEORY

17.10. KINETIC THEORY– UNDERPINNINGS

Do You Believe in Atoms?

Some of our ancestors believed in the reality of witches. In fact, they thought that they had good evidence that witches existed, good enough evidence to accuse some people of being witches. We believe in atoms. Are we truly more scientific than they were?

17.10.1. Activity: Why Atoms!?

a. Describe several reasons why you do or do not believe that matter consists of atoms and molecules, even though you have never seen them with your own eyes.

b. What happens when thermal energy is transferred to a substance? If you believe that substances are made of atoms and molecules, how might you explain the change in volume of a heated gas?

Models of Pressure Exerted by Molecules

So far in physics we have talked about matter as if it were continuous. We didn't need to invent aluminum atoms to understand how a ball rolled down the track. As early as the fifth century B.C.E. Greek philosophers, such as Leucippus and Democritus, proposed the idea of "atomism." They pictured a universe in which everything is made up of tiny "eternal" and "incorruptible" particles, separated by "a void." Today, we think of these particles as atoms and molecules.

In terms of everyday experience, molecules and atoms are hypothetical entities. In just the past forty years or so, scientists have been able to "see" molecules using electron microscopes and field ion microscopes. But long before atoms and molecules could be "seen," nineteenth-century scientists such as James Clerk Maxwell and Ludwig Boltzmann in Europe and Josiah Willard Gibbs in the United States used these imaginary *microscopic* entities to construct models that made the description and prediction of the *macroscopic* behavior of thermodynamic systems possible. Is it possible to describe the behavior of an ideal gas that obeys the first law of thermodynamics as a collection of moving molecules? To answer this question, let's observe the pressure exerted by a hypothetical molecule undergoing elastic collisions with the walls of a two-dimensional box. By using the laws of mechanics we can derive a mathematical expression for the pressure exerted by the molecule as a function of the volume of the box. If we then define temperature as being related to the average kinetic energy of the molecules in an ideal gas, we can show that kinetic theory is compatible with the ideal gas law and the first law of thermodynamics. This compatibility doesn't prove that molecules exist, but allows us to say that the molecular model enables us to explain aspects of the experimentally determined ideal gas law.

17.11. 2D MOLECULAR MOTION AND PRESSURE

Consider a spherical gas atom that has velocity $\vec{v} = v_x \hat{x} + v_y \hat{y}$ and makes perfectly elastic collisions with the walls of a two-dimensional box of length L_x and width L_y as shown in Figure 17.8.

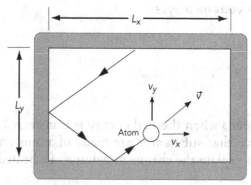

Fig. 17.8. A typical screen image of a single atom moving at a speed, v, in a two-dimensional box. Assume the atom bounces off the walls of the box with perfectly elastic collisions.

You will be making measurements on the simulated motions of an atom bouncing in a box to derive the mathematical relationship that relates pressure of the gas, P, to the kinetic energy of the atom and the volume, V, occupied by the gas. To complete this exercise you will need:

- 1 molecular motion computer simulation (of a single atom bouncing around in a 2D box) such as M171101(AtomInBox).mov or the PRU001.mov

EITHER

- 1 stopwatch
- 1 ruler

OR

- video analysis software

Recommended group size:	2	Interactive demo OK?:	N

In the next activity, you will use the hypothesis that a gas is made of a large collection of atoms behaving like little billiard balls to explain why the ideal gas law might hold. In the next activity you are to pretend you are looking under a giant microscope at a single spherical atom as it bounces around in a two-dimensional box by means of elastic collisions and that you can time its motion and measure the distances it moves as a function of time.

If the atom obeys Newton's laws, you can calculate how the average pressure, ΔP_x, that the atom exerts on the walls of its container is related to the volume, V, of the box. What is the momentum change as the atom bounces off a wall in terms of the change in the velocity component, v_x, perpendicular to the wall? How often will our atom "hit the wall" as a function of its component of velocity perpendicular to the wall and the distance between opposite walls? What happens when the atom is more energetic and moves even faster? Will the results of your calculations based on mechanics be compatible with the ideal gas law?

Note: Remember that capital P represents pressure while lower case p_x represents a momentum component. Similarly, V, denotes volume where v_x represents a velocity component.

17.11.1. Activity: The Theory of 2D Molecular Motion

a. Suppose the molecule moves across a distance L_x completely across the box in the x-direction in a time Δt_x. What is the equation needed to calculate its x-component of velocity in terms of L_x and Δt_x?

b. Suppose the molecule moves a distance L_y completely across the box in the y-direction in a time Δt_y. What is the equation needed to calculate its y-component of velocity in terms of L_y and Δt_y?

c. Measure the dimensions L_x and L_y of the box on your computer screen.

d. Use a ruler or video analysis software to measure the average time, Δt_x, it takes the atom to move from the left wall of the box to the right wall. Also determine Δt_y.

e. Calculate the magnitude of the x and y components of velocity.

f. Use a stopwatch and ruler or video analysis software to measure the total length of the straight path of the molecule in centimeters as it moves between a collision with the right wall and a collision with the bottom wall. Also determine the time between the collisions. Use your data to compute the speed of the molecule.

g. How does the measured speed of your molecule compare to that determined by the equation $v^{total} = \sqrt{(v_x^2 + v_y^2)}$?

h. Suppose the box were a three-dimensional container. Write the expression for v^{total} in terms of the x, y, and z components of velocity. **Hint:** This is an application of the 3-dimensional Pythagorean theorem.

i. We would like to find the average kinetic energy of each molecule. Since the kinetic energy of a molecule is proportional to the square of its total speed, you need to show that if *on the average* $v_x^2 = v_y^2 = v_z^2$, then $(v^{total})^2 = 3v_x^2$.

j. Assume that the interaction time between the molecule and the wall is negligible. If the molecule bounces back at the same speed in the x-direction, show that the time, Δt, it takes to return to the left wall is twice the transit time from one wall to the other and back.

k. If the collisions with the wall perpendicular to the x-direction are elastic, show that the x-component force exerted on that wall for each collision is just $F_x = 2m(v_x)/(2\Delta t_x)$ where $2\Delta t_x$ is the transit time and m is the mass of the particles and $2\Delta t_x$ the mean interval between collisions with one of the walls. (**Hint:** Think of the form of Newton's second law in which force is defined in terms of the change in momentum per unit time so that $F_x = \Delta p_x/(\Delta t.)$ **Warning:** Physicists often use the same symbol to stand for more than one quantity. In this case, note that Δp (where "p" is in lowercase) indicates the change in momentum, and not pressure change.

l. Substitute the expression from part a. for Δt_x to show that

$$F_x = \frac{m v_x^2}{V}$$

m. For simplicity, let us assume that we have a cubical box so that the length, width, and height are all the same so that $L = L_x = L_y = L_z$. Hence, the volume of the box is the cube of its length. In other words, $V = L^3$. Show that the pressure on the wall perpendicular to the x-axis caused by the force F_x due to one atom is

$$P = \frac{m v_x^2}{V}$$

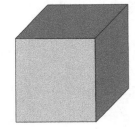

Fig. 17.9. The volume of the box is $V = L_x L_y L_z = L^3$.

n. Let's say that there are not one but N molecules in the box. What is the pressure on the wall now?

Note: For simplicity we use a cubical box. You should be able to convince yourself that you will obtain the same result for a box of dimensions L_x, L_y, and L_z.

o. Next, show that if we write the volume of our box as $V = L^3$, and recalling that $v_x^2 = \dfrac{(v^{total})^2}{3}$, we can write $P = N\dfrac{m(v^{total})^2}{3V}$.

p. Finally, since the average kinetic energy of a molecule is just $\langle K \rangle = \frac{1}{2}m(v^{total})^2$, show that the pressure in the box can be written as

$$P = \frac{2N\langle K \rangle}{3V}$$

17.12. KINETIC ENERGY, INTERNAL ENERGY, AND TEMPERATURE

The Effect of Increasing Volume and the Number of Atoms

We have hypothesized the existence of non-interacting atoms to provide the basis for a particle model of ideal gas behavior. We have shown that the pressure of such a gas can be related to the average translational kinetic energy of each atom:

$$P = \frac{2N\langle K \rangle}{3V} \qquad \text{or} \qquad PV = \tfrac{2}{3}\, N\langle K \rangle$$

Pressure increases with kinetic energy per atom and decreases with volume. This result makes intuitive sense. The more energetic the motions of the molecules, the more pressure we would expect them to exert on the walls. Increasing the volume of the box decreases the frequency of collisions with the walls, since the molecules will have to travel longer before reaching them, so increasing volume should decrease pressure if $\langle K \rangle$ stays the same.

17.12.1. Activity: Gas Law and Kinetic Theory Predictions

a. According to the ideal gas law, $PV = nRT = Nk_BT$. What should happen to the pressure of an ideal gas as its volume increases? As the number of particles increases?

b. What do you predict will happen to the pressure in your simulated gas if you increase the volume of its container? Explain your reasoning.

c. What do you predict will happen to the pressure in your simulated gas, if you increase the number of atoms in the box?

Kinetic Theory and the Definition of Temperature

The model of an ideal gas we have just derived requires that:

$$PV = \frac{2}{3} N\langle K \rangle$$

But according to Equation 17.4, one form of the ideal gas law is given by

$$PV = Nk_B T$$

where N = the total number of gas molecules in the gas

and k_B = Boltzmann's Constant given by 1.38×10^{-23} J/K

What can we say about the average translational kinetic energy per molecule for an ideal gas? You can derive a relationship between temperature and the energy of molecules that serves as a microscopic or molecular definition of temperature.

17.12.2. Activity: Microscopic Definition of T

a. Use the ideal gas law and the equation relating N, P, and V to the translational kinetic energy of an atom that you just developed to derive an expression relating $\langle K \rangle$ and T. Show the steps in your derivation.

b. In general, a gas that is made up of molecules can store internal energy by rotating or vibrating. For an ideal gas of point particles like the atoms we have considered, the only possible form of internal energy is the sum of the kinetic energies of the atoms. If we can ignore potential energy due to gravity or electrical forces, then the internal energy, E^{int}, of a gas of N particles is $E^{int} = N\langle K \rangle$. Use this to show that for an ideal gas of point particles, E^{int} *depends only on N and T.* Derive the equation that relates E^{int}, N, and T for an ideal gas. Show the steps.

The microscopic and the macroscopic definitions of temperature are equivalent. The microscopic definition of temperature that you just derived is fundamental to the understanding of all of thermodynamics!

A Note About Internal Energy

The microscopic model for an ideal gas is that of a collection of point particles that undergo perfectly elastic collisions. We noted that the internal energy in an ideal gas is just the sum of the translational kinetic energies of its particles.

The molecules in a real gas often have structures that can rotate and vibrate. Thus, several types of energy must be taken into account in determining internal energy.

In *liquids and solids* atoms and molecules exert forces on each other and are part of a complex system that has both potential and kinetic energy.

UNIT 18: HEAT ENGINES

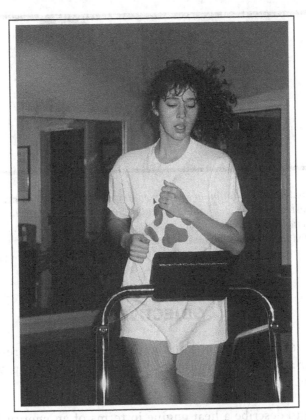

In general, engines convert various forms of energy into mechanical work. For example, this athlete uses chemical energy released during the oxidation of molecules obtained from food to do mechanical work. The efficiency of her muscles in transforming chemical energy into work is at best only about 20%. Thus, 80% of the chemical energy released during physical activity is ultimately transferred elsewhere as thermal energy. Since this athlete must maintain her internal body temperature at about 37 degrees Celsius, this thermal energy must be transferred to her surroundings.

The nineteenth-century industrial revolution was based on the invention of heat engines. Heat engines have much in common with the chemical engines that power humans. For instance, both human engines and heat engines extract energy at a higher temperature, do work, and transfer thermal energy to lower temperature surroundings. Even though the 20% efficiency of a human engine seems low, it is far higher than the efficiency of a heat engine working between the same two temperatures. When you complete this unit, you will have a better understanding of how the laws of thermodynamics allow us to place absolute limits on the efficiency of heat engines.

UNIT 18: HEAT ENGINES

The production of motion in a steam engine is always accompanied by a circumstance which we should particularly notice. This circumstance is the passage of caloric from one body where the temperature is . . . elevated to another where it is lower.

S. N. L. Carnot (1824)

OBJECTIVES

1. To be able to derive relationships between temperature, pressure, and volume for adiabatic and isothermal expansions and compressions of an ideal gas.

2. To be able to describe a heat engine in terms of an energy flow diagram and to calculate the work done in a cycle.

3. To investigate, both theoretically and experimentally, the relationship between work done by a heat engine and changes in the pressure and volume of the engine's working medium.

4. To explore theoretical aspects of the Carnot engine, which is designed to be the most efficient possible heat engine.

18.1. OVERVIEW

Often scientific developments are closely related to the demands of a new technology. Historically, the development of the science of thermodynamics, and in particular of the thermodynamics of gases, was motivated by the desire to build better heat engines. The steam engine, which issued in the industrial revolution, and later, the internal combustion engine, both depend on "cycles" in which gases are alternately expanded and then compressed. The end result of these cycles is that a portion of thermal energy transferred to gas is converted into work. An understanding of the detailed physics of the expansion and compression of gases has helped engineers to design more efficient engines.

For simplicity, we will discuss the expansion and compression of ideal gases in this unit. We will also look at the more general idea of a "cycle" that can convert some of the thermal energy absorbed by a system to work.

An ideal engine would transform 100% of the thermal energy it absorbs to useful work. But, both real experience and the work of Sadi Carnot in the 19th century show that it is impossible to design an engine that is 100% efficient. Finally, you will analyze the operation of the Carnot engine which is believed to be the most efficient of all conceivable engines. And, you will calculate the thermal efficiency of a Carnot engine cycle involving compressions and expansions of an ideal gas.

COMPRESSION AND EXPANSION OF GASES

18.2. DOES THE IDEAL GAS LAW TELL ALL?

Since many practical devices, from automobile engines to refrigerators, use expanding gases to operate, it is important to understand what happens to gases when they undergo volume changes. We know that $PV = nRT$ for an ideal gas. Can this relationship used by itself tell us what happens to the temperature of a gas if its volume changes? Discuss your answer with your partner.

18.2.1. Activity: Ideal Gas Tells All?

a. Can the ideal gas law be used to calculate the change in temperature of a system as its volume increases? Why or why not?

18.3. ISOTHERMAL AND ADIABATIC PROCESSES FOR AN IDEAL GAS

Imagine a piston filled with an ideal gas. What could happen to it as it is compressed? The first law of thermodynamics tells us that

$$\Delta E^{\text{int}} = Q - W = Q - \int P dV \qquad (18.1)$$

If we are going to describe what happens when the volume of a gas changes in more detail, we might try using the ideal gas law in conjunction with the first law of thermodynamics. Let's start by considering an isothermal process in which there is no temperature change in a gas while it is being compressed.

18.3.1. Activity: Isothermal Compression of a Gas

a. In an earlier section you showed that, for an ideal gas, $E^{\text{int}} = (3/2) Nk_BT$. Now show that, for a fixed number of molecules of gas, N, a change in the temperature of an ideal gas ΔT produces a change in internal energy that is proportional to ΔT and given by $\Delta E^{\text{int}} = (3/2) Nk_B\Delta T = (3/2)nR\Delta T$.

b. In an *isothermal* compression, the gas is kept at a constant temperature by thermal energy transfer to its surroundings. Therefore ΔT and ΔE^{int} are both zero in an isothermal compression. Find an expression relating Q and W and another expression relating P and V during an isothermal compression. **Hint:** Use both the first law of thermodynamics and the ideal gas law.

c. The Boyle's law experiment that some of you carried out in an earlier unit was in fact an isothermal compression, since the syringe was always in thermal equilibrium with the room. Are the Boyle's law measurements you or your classmates made consistent with what you derived above? Explain.

Another type of process that can occur during the expansion or compression of a gas is an *adiabatic* change. An adiabatic process is defined as one in which a system does not exchange thermal energy with its surroundings so that $Q = 0$ during the process. This can be brought about either by carefully insulating the system so that no thermal energy exchange is possible, or by carrying out the process so rapidly that thermal or heat energy transfer does not have time to take place. What happens to an ideal gas if it is compressed adiabatically? We would like you to show on a step-by-step basis, outlined in Activity 18.3.2 through Activity 18.4.2 below, that for an ideal gas undergoing an adiabatic expansion the following expression can be used to describe the relationships between an initial volume and temperature and a final volume and temperature.

$$\frac{3}{2}\frac{dT}{T} + \frac{dV}{V} = 0 \quad \text{so that} \quad T_2^{3/2} V_2 = T_1^{3/2} V_1$$

Note: The exponent of 3/2 only holds for an ideal monatomic gas. For a "real" gas the exponent will be different.

18.3.2. Activity: Adiabatic Compression of a Gas

a. The first step: In Activity 18.3.1 you showed that the change in the internal energy of an ideal gas is given by $\Delta E^{int} = (3/2)Nk_B\Delta T$. Use the first law of thermodynamics to find a relationship between the work done when an ideal gas is compressed adiabatically by an amount ΔV (with no change in pressure) and the change in the temperature of the gas, ΔT. In particular, show that $(3/2)Nk_B\Delta T = -P\Delta V$. **Hint:** For a small change in volume at constant pressure the amount of work done, ΔW, is given by $\Delta W \approx P\Delta V$.

b. Next you can use the ideal gas law and the relationship you just derived to show that for small temperature changes the fractional change in the temperature of an ideal gas, $\Delta T/T$, can be related to the fractional change in volume, $\Delta V/V$, by

$$\frac{3}{2}\frac{\Delta T}{T} = -\frac{\Delta V}{V} \quad \text{or} \quad \frac{3}{2}\frac{\Delta T}{T} + \frac{\Delta V}{V} = 0$$

18.4. MATHEMATICAL INTERLUDE: LN(X) AND ∫DX/X

In the study of adiabatic processes we sometimes encounter equations with some thermodynamic variable raised to a power, for example, P^γ. It is often useful in manipulating such equations to take logarithms. Some of you may already be familiar with common or base 10 logarithms. Usually denoted $\log(x)$ or $\log_{10}(x)$, the logarithm of x is that power to which the base number of 10 must be raised in order to obtain x. For example, 10 must be raised to the third power to get 1000, since $10^3 = 1000$; consequently $\log_{10}(1000) = 3$. Put in terms of symbols, if $x = 10^r$ then $\log_{10}(x) = r$. Logarithms of numbers follow certain rules that make them helpful calculation aids. For example, the relations:

$$\log_{10}(A \times B) = \log_{10}(A) + \log_{10}(B)$$
$$\log_{10}(A/B) = \log_{10}(A) - \log_{10}(B)$$
$$\log_{10}(A^n) = n\log_{10}(A)$$

hold true for any positive numbers A and B and exponent n.

It is possible to use a base other than 10 to set up a system of logarithms. Frequently in mathematics, base e logarithms $(\ln(x))$ are used, where $e = 2.7182818294\ldots$ and is a transcendental number. For base e or *natural* logarithms, if $x = e^y$ then $\ln(x) = y$. All the rules for $\log_{10}(x)$ listed above work for $\ln(x)$. We are interested in the derivative of $\ln(x)$ as it comes up frequently in physics. It is also the case that $\ln(x)$ is the integral of a useful, simple function.

18.4.1. Activity: Derivative of ln(x)

Here is a graph of $\ln(x)$. Notice that it has a very steep slope near $x = 1$ but a much more gradual slope near $x = 16$.

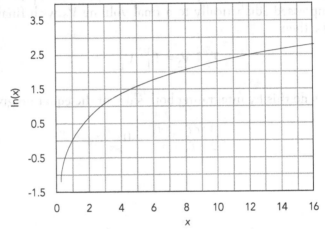

a. By inspecting this graph, where do you expect the slope of $\ln(x)$ (that is, its derivative) to be large? Where is it small?

b. Look up the solution for the derivative of the function $y = \ln(x)$ in the appendix of an introductory physics or math text and write it down in the space below.

$$\frac{d[\ln(x)]}{dx} =$$

c. Use the derivative in part b above to express the integral $\int dx/x$. (Recall that the integral of a function is its anti-derivative.) Or use the appendix of an introductory physics or math text to determine $\int dx/x$. In either case, write it down in the space below.

$$\int_{x_1}^{x_2} \frac{dx}{x} =$$

d. Using the result of part c. above and substituting the initial and final volume of a gas as the limits of integration V_1 and V_2, evaluate the integral shown below where a gas volume V plays the role of the independent variable x.

$$\int_{V_1}^{V_2} \frac{dV}{V} =$$

Now we are back to proving that

$$T_2 V_2^{3/2} = T_1^{3/2} V_i \tag{18.2}$$

18.4.2. Activity: *T* vs. *V* for Adiabatic Expansions

Combine the results you just obtained in Activities 18.3.2b and 18.4.1c to show that, if a gas of initial temperature T_1 and volume V_1 is compressed adiabatically to a final volume V_2 with final temperature T_2, then

$$T_2^{3/2} V_2 = T_1^{3/2} V_1$$

You can do this by integrating both sides of the equation given by

$$\frac{3}{2} \frac{dT}{T} + \frac{dV}{V} = 0$$

Note: Strictly speaking, this result only holds for ideal monatomic gases composed of single molecules such as helium. A similar equation holds for other gases at modest densities.

Fig. 18.1.

18.5. THE FIRE SYRINGE AND THE RAPID COMPRESSION OF AIR

A device known as a *fire syringe* allows a rapid compression of air in a small glass tube that is inside a safety tube of Plexiglas. If pushed very hard, the piston in the glass tube can be forced almost down to the end of the straight-walled section of the tube. If this is done rapidly, the compression can be nearly adiabatic. Air is not a monatomic gas, but the formulas derived above work well enough. As you can tell from the equation you derived in the last activity, the air in the fire syringe should increase in temperature as its volume decreases. Examine a fire syringe and make some reasonable assumptions about the initial and final volumes of air in the chamber. You can then calculate the approximate final temperature of the compressed air. Finally, you can attempt to ignite a tiny piece of tissue paper with the fire syringe. For this activity you will need:

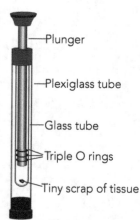

- 1 fire syringe
- 1 ruler
- 1 glass thermometer
- 1 piece of tissue paper, approx. 0.5 mm on a side

Recommended group size:	2	Interactive demo OK?:	Y

Fig. 18.2. A fire syringe that allows a rapid compression of trapped air to ignite a piece of paper.

18.5.1. Activity: The Fire Syringe–Fahrenheit 451

a. Before approximating the final temperature of air in the syringe, estimate the following quantities:

Initial length of air column L_1 = _____ cm

Final length of air column L_2 = _____ cm

Inner radius of the tube R = _____ cm

Initial volume V_1 = _____ cm^3

Final volume V_2 = _____ cm^3

Initial temperature T_1 = _____ K

b. Calculate the final temperature in the cylinder in Kelvin.

Final Temperature T_2 = _____ K

c. Compare this to the "flash point" or burning temperature of paper, which is 451°F.* What do you expect to happen to the tissue paper in the fire syringe when the plunger is pushed down rapidly?

d. Put on safety gloves and carry out the fire syringe experiment by rapidly and forcefully depressing the plunger. What happens?

e. Why doesn't the tissue paper catch fire when you compress the air slowly?

---

* The paper flashpoint of 451 degrees Fahrenheit is well known to readers of Ray Bradbury's famous science fiction novel, *Fahrenheit 451*, about book burning.

We have examined in detail an isothermal and an adiabatic compression. Other processes are possible: A process with no pressure change is called *isobaric* and one with no volume change is called *isovolumetric*.

HEAT ENGINES AND CYCLES

18.6. OVERVIEW OF HEAT ENGINES

As you may already know, the internal combustion engines inside cars and trucks are heat engines. So is the steam engine of that bygone era of railroading that led to the spanning of the American continent. The word "engine" conjures up an image of something that we start up, that runs, and that provides a continuous flow of work. For example, a car engine accelerates us down the road and when we have reached a (safe!) speed, the engine helps us maintain the car's speed by overcoming air drag.

Fig. 18.3. An internal combustion heat engine hidden inside a van.

The basic goal of any heat engine is to absorb thermal energy and transform it into useful work as efficiently as possible. This is done by taking a *working medium*, some substance that can expand or contract and thus do work after it absorbs thermal energy. But the working medium must be part of a system designed to produce work in continuous, repeated cycles.

An engine that is one-hundred percent efficient would have a working medium that absorbs thermal energy from a high temperature entity and transforms all of it to useful work. Such a process does not violate the first law of thermodynamics. However, no heat engine has ever been capable of transforming *all* of the thermal energy transferred to it into useful work. Some of the absorbed thermal energy is always transferred back to the engine's surroundings. Furthermore, the thermal energy transfer to the surroundings takes place at a lower temperature and is thus less capable of being transformed to useful work. The inevitable presence of thermal energy transfer back to the environment has led scientists to formulate a second law of thermodynamics. A common statement of the second law is simply that it is impossible to transform all of the thermal energy transferred to a system into useful work.

In the rest of the activities in this unit you will learn more about the behavior of ideal gases as they are expanded and compressed. You will explore the actual behavior of several simple heat engines. Finally, you will consider the limitations in the efficiency of real engines in accordance with the first and second laws of thermodynamics.

18.7. THE RUBBER BAND AS A HEAT ENGINE MEDIUM

We can use any substance that changes its volume when thermal energy is transferred to it as the working medium for a heat engine. The working medium is capable of transforming a fraction of the thermal energy absorbed by it into useful work. Both rubber bands and gases can be used as working media. In the activities in this section you will explore the characteristics of heat engines using a rubber band as the working medium. In the sections that follow you will study the compression and expansion of gases used as working substances in heat engines. As you complete this unit, we hope you will begin to understand that there are general principles governing the operation of

Fig. 18.4. When gasoline is burned inside of a internal combustion engine, thermal energy is transferred to the air in cylinders inside the engine.

heat engines that do not depend on the detailed nature of the *working medium*.

Lifting a Mass with a Rubber Band

Let's examine what happens when a large rubber band that is stretched by a hanging weight is heated with a heat gun or hair dryer. The equipment needed for this demonstration is as follows:

- 1 large rubber band, 5 cm × 10 cm
- 1 mass pan
- 2 masses, 10 kg
- 1 table clamp
- 1 long rod
- 1 short rod
- 1 right angle clamp
- 1 heat gun (or 1500 W hair dryer with Styrofoam housing around rubber band to keep heat from being transferred to the surrounding)

Recommended group size:	All	Interactive demo OK?:	Y

18.7.1. Activity: The Working Rubber Band

a. What do you predict will happen when the rubber band is heated?

b. What actually happens?

c. How does the behavior of the heated rubber band compare with that of a piston in a gas cylinder when the gas is heated? What are the similarities and differences?

Developing a Mass Lifting Engine Cycle

This behavior of a heated rubber band is fine if all we want to do is to move a weight once, but it hardly fits our intuitive notion of an engine. Let's say you own a factory that produces a canned beverage. (We will let you choose the beverage!) You wish to design a machine that lifts cans from the conveyor belt leaving the filling machine up to the conveyor belt entering the packing machine. You have at your disposal a large rubber band and a hair dryer. Can you design a mass lifter? It might look like the one in the following diagram.

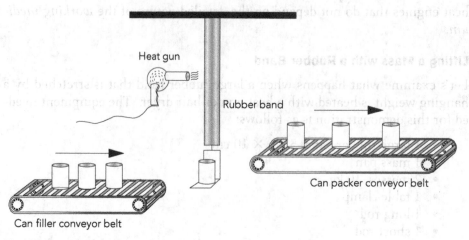

Fig. 18.5.

18.7.2. Activity: Can Lifter Can-Do!

a. What do you need to do to lift the first can to the top of the can-packing conveyor belt?

b. What do you need to do to the can once it has reached the packing conveyor belt?

c. What must happen so that the lifter will be ready to pick up another can at the production conveyor belt? What has to happen to the rubber band?

d. Can you use your answers above to describe a *cycle* that would repeatedly lift cans to the packing conveyor? Describe the steps in your cycle.

e. Carefully point out where heat or thermal energy is transferred in your cycle, the direction of thermal energy transfer, where work is done, and by which part of the engine that work is done.

    f. What would happen to the engine on a very hot day when the temperature inside the factory was as hot as the heated air coming from the hair dryer or heat gun?

    g. Suppose that all the thermal energy given off by the hair dryer is absorbed by the rubber band. Does it appear likely that our rubber band lifter converts *all* of the thermal energy it absorbs from the hair dryer into useful mechanical work (that is, work done lifting cans)? Does any of that absorbed thermal energy have to go elsewhere? **Hint:** What has to happen to the rubber band after the lifted can is taken away but before it can pick up a new can from the lower belt?

18.8. ENERGY FLOW DIAGRAMS, CYCLES, AND EFFICIENCY

One of the key features of our rubber band engine is that to be ready to lift the next can, the rubber band must cool by giving off thermal energy to its surroundings, which are colder than the hot air from the hair dryer. After the rubber band has cooled and stretched back to its original shape, it is in the same *thermodynamic state* that it was in at the start; that is, all its properties, including its internal energy, E^{int}, are the same. *For one complete cycle of our rubber band engine $\Delta E^{int} = 0$ J.*

    If Q_H is the thermal energy absorbed by the rubber band from the air heated by the hair dryer and Q_C is the thermal energy transferred to the cooler room air, the net thermal energy absorbed in the cycle is $Q = Q_H - Q_C$ and the first law of thermodynamics becomes

$$\Delta E^{int} = Q - W = (Q_H - Q_C) - W \tag{18.3}$$

Since $\Delta E^{int} = 0$ J for our *complete* cycle, we can simplify this by writing:

$$W = Q_H - Q_C$$

The Carnot Engine

As noted by Sadi Carnot in 1824, all heat engines have this characteristic: the performance of useful work is accompanied by thermal energy being transferred to the working medium from a hot body and some thermal energy subsequently being transferred to a cooler body. The difference between these thermal energies constitutes the useful work that can be done by the cycle. The hot body is usually called the *high temperature reservoir* and the cool body is the *low temperature reservoir*. The word reservoir is used because we imagine that we have so much matter that we can transfer an amount of thermal energy, Q_H, from the hot reservoir or dump thermal energy, Q_C, to the cold reservoir without changing their temperatures measurably.

This basic fact about heat engines is often discussed in terms of an energy flow diagram such as the one shown below. This diagram would work equally well for an old-fashioned steam engine or our rubber band can lifter.

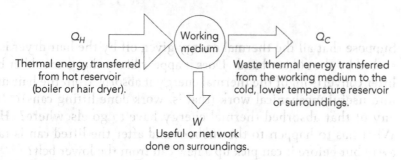

Fig. 18.6. Heat engine schematic.

Figure 18.6 is a pictorial representation of what we have written in words: our engine has thermal energy Q_H transferred to it, does work W, and transfers some of the original thermal energy Q_C to lower temperature surroundings.

We can define the efficiency of any engine cycle as the desired amount of work produced as a result of the cycle (in this case the useful or net work done) divided by the magnitude of the thermal energy that must be transferred to the working medium to achieve the result (in this case the $|Q_H|$ absorbed from the hair dryer). The efficiency of a heat engine is usually denoted as η ("eta"). In defining efficiency physicists often put bars around the symbols for thermal energy and work to designate that it is the magnitude of energy being transferred in or out of the system that is to be used in equations for efficiency.

18.8.1. Activity: Defining Efficiency

a. Use the definition of efficiency to write the equation for engine efficiency, η, in terms of the magnitude of the net work done, $|W|$, and the magnitude of the thermal energy transferred to the working medium from the hot reservoir, $|Q_H|$.

b. Use the first law of thermodynamics and the fact that after completion of a full engine cycle the net change in the internal energy of the working medium is zero to show that

$$\eta = 1 - \frac{|Q_C|}{|Q_H|}$$

(18.4)

Hint: $W \equiv |Q_H| - |Q_C|$ where $|Q_C|$ is the magnitude of the waste thermal energy transferred to the cold reservoir during the cycle.

V increasing
P constant

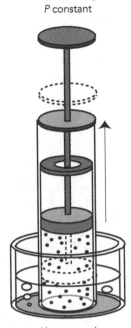

Hot reservoir

Fig. 18.7. Part 1 with cylinder and piston in hot water.

c. If it were possible to design an engine with no waste thermal energy being dumped into a cold reservoir, what would the efficiency of the engine be?

18.9. A HEAT ENGINE USING A SIMPLE GAS CYCLE

A Theoretical Analysis of a Heat Engine Cycle

To get a better idea of how real heat engines use gases as a working medium, let's analyze a cycle made up of changes in volume made under constant pressure (isobaric) and changes in pressure made under constant volume (isovolumetric). Although this cycle is not realistic, it is easy to analyze and bears some resemblance to the cycles used in the internal combustion engine.

We start with a fraction of a mole of an ideal gas that is initially at room temperature. If the pressure of the piston, rod, and platform are small compared to atmospheric pressure, then the gas pressure will be very close to atmospheric pressure. For example, suppose the gas has an initial pressure of 1.02×10^5 N/m^2 and an initial volume of 0.08 m^3. Call this situation "Point A." In part 1 of the cycle the cylinder is placed in hot water. The hot water transfers just enough thermal energy to the gas so that it expands at *constant* pressure and does work on its surroundings until it has reached a volume of 0.10 m^3; call this situation "Point B."

Next, in part 2, the gas is placed in a cold reservoir and is cooled at a constant volume of 0.10 m^3. This can be done by placing a collar between the piston and the platform so the piston cannot descend. Suppose the gas transfers thermal energy to the ice water until its pressure decreases to 0.79×10^5 N/m^2. Call this "Point C."

In part 3, we remove the collar and let gas cool further. At the same time we partially support the piston to keep the gas at a lower constant pressure of 0.79×10^5 N/m^2 while its volume is reduced to the original 0.08 m^3. In part 4 we clamp the piston, remove it from the water, and allow the pressure on the gas to rise to its original 1.02×10^5 N/m^2. Thus, we have returned to "Point A" again.

18.9.1. Activity: Analyzing the Cycle

a. Complete the plot of the cycle just described in the *P-V* diagram that follows by adding parts 3 and 4 to the cycle. Label each part.

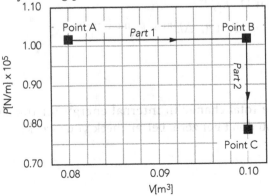

V constant
P decreasing

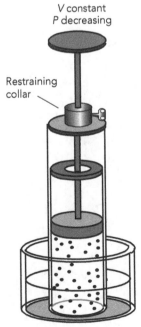

Restraining collar

Cold reservoir

Fig. 18.8. Part 2 with cylinder and piston placed in ice water.

b. During which parts of the cycle is work done by the gas? On the gas? Explain.

c. During which parts of the cycle is thermal energy transferred to the gas from a reservoir? From the gas to a reservoir? Explain.

d. Use the fact that $E^{int} = (3/2)nRT$ and $PV = nRT$ so that $E^{int} = (3/2)PV$ to evaluate E^{int} at each of the four "points" of the cycle. Then evaluate the change in internal energy ΔE^{int} for each part of the cycle.

Point A: $E_A{}^{int} =$ _____

Part 1: $\Delta E^{int} = E_B{}^{int} - E_A{}^{int} =$ _____

Point B: $E_B{}^{int} =$ _____

Part 2: $\Delta E^{int} = E_C{}^{int} - E_B{}^{int} =$ _____

Point C: $E_C{}^{int} =$ _____

Part 3: $\Delta E^{int} = E_D{}^{int} - E_C{}^{int} =$ _____

Point D: $E_D{}^{int} =$ _____

Part 4: $\Delta E^{int} = E_A{}^{int} - E_D{}^{int} =$ _____

e. What is the total change in internal energy for the complete cycle? **Hint:** If you don't get zero, better check your work!

f. List the change in internal energy ΔE^{int} and calculate the work done $(P\Delta V)$ and the thermal energy exchanged (from the first law of thermodynamics) for each part of the cycle.

1: ΔE^{int} = _____ Thermal Energy Transferred = _____ Work = _____

2: ΔE^{int} = _____ Thermal Energy Transferred = _____ Work = _____

3: ΔE^{int} = _____ Thermal Energy Transferred = _____ Work = _____

4: ΔE^{int} = _____ Thermal Energy Transferred = _____ Work = _____

g. Explain why the work in part 1 of the cycle in which the gas is expanding is positive while the work in part 3 of the cycle in which the gas is contracting is negative.

h. The expanding gas in part 1 of the cycle does work on its surroundings. Alternatively the surroundings do work on the gas to compress it in part 3 of the cycle. The net work done on the surroundings in an engine cycle is defined as the sum of the positive work done on the surroundings and the negative work done by the surroundings. Calculate the net work done on the surroundings by the gas as it expands and contracts during a cycle using the equation

$$W^{net} = W_1 + W_2 + W_3 + W_4$$

$W^{net} =$

i. Show that the net work you just calculated is approximately the same as the area enclosed by the rectangle determined by points A, B, C, and D on the P-V diagram that you filled out in part a. of this activity.

Finding Net Work Using a P-V Cycle

In general, when a gas is expanding it is doing positive work *on* the surroundings and when it is being compressed work is being done on it *by* the surroundings. When the work done *by* the surroundings on a gas is calculated properly, it comes out to be negative.

Typically, at the completion of a heat engine cycle, the gas has the same internal energy, temperature, pressure, and volume that it started with. It is then ready to begin another cycle. During various phases of the cycle the thermal energy transferred to the gas from the hot reservoir causes: (1) the gas to do work on its surroundings as it expands, (2) the surroundings to do work on the gas to compress it, and (3) the gas to transfer waste thermal energy to the surroundings or cold reservoir.

Real heat engines have linkages between a moving piston and the gas or other working medium that allows the expansion and compression phases of the cycle to run automatically. Thus, some of the work done on the surroundings provides the work energy needed to compress the gas to return it to its starting point. The useful or net work done in an engine cycle must account for the positive work done during expansion and the negative work done during compression. Theoretically, the net work can be calculated by taking the integral of PdV around a complete cycle using the expression

$$W^{\text{net}} = \oint PdV \qquad (18.5)$$

Note: The symbol $\oint$ represents an integration around a complete cycle.

It can be shown mathematically that this integral around a closed loop is the same as the area enclosed by the trace of the combinations of pressure and volume on a $P\text{-}V$ diagram during a complete cycle. This is illustrated in Figure 18.9. For example, you should have found that the area enclosed by the diagram in Activity 18.9.1a is the same as the net work you calculated in part f. of the activity.

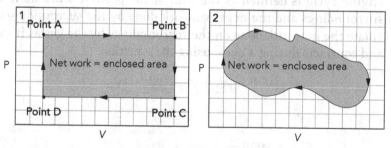

Fig. 18.9. (1) The enclosed area on a $P\text{-}V$ diagram for the simple gas engine cycle introduced in this section; (2) the enclosed area for an arbitrary, more complex engine cycle.

In the next activity, you will attempt to verify this relationship between useful work and the area on a $P\text{-}V$ diagram for a real engine.

THE INCREDIBLE MASS LIFTING MACHINE

18.10. LAB EXPERIMENT – THE MASS LIFTER HEAT ENGINE

Your working group has been approached by the Newton Apple Company about testing a heat engine that lifts apples that vary in mass from 100 g to

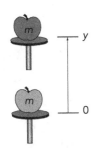

Fig. 18.10. Doing useful mechanical work by lifting a mass, *m*, through a height, *y*.

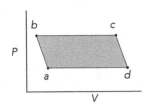

Fig. 18.11. Doing thermodynamic work in a heat engine cycle.

200 g from a processing conveyer belt to the packing conveyer belt that is 10 cm higher. The engine you are to experiment with is a "real" thermal engine that can be taken through a four-stage expansion and compression cycle and that can do useful mechanical work by lifting small masses from one height to another. In this experiment we would like you to verify experimentally that the useful mechanical work done in lifting a mass, *m*, through a vertical distance, *y*, is equal to the net thermodynamic work done during a cycle as determined by finding the enclosed area on a *P-V* diagram. Essentially you are comparing useful mechanical "*mgy*" work (which we hope you believe in and understand from earlier units) with the accounting of work in an engine cycle as a function of pressure and volume changes given by the expression in Equation 18.5.

Although you can prove mathematically that this relationship holds, the experimental verification will allow you to become familiar with the operation of a real heat engine. To carry out this experiment you will need:

- 1 syringe, 10 cc*
- 1 length of Tygon® tubing, 30 cm (1/8" ID)
- 1 Erlenmeyer flask, 25 ml
- 1 #0 1-hole rubber stopper
- 1 rod stand
- 1 rod
- 1 test tube clamp
- 2 coffee mugs (to use as reservoirs)
- 1 50 g mass
- crushed ice, approx. 50 ml
- small tray (to prevent spilling)
- 1 electronic scale

  OPTIONAL

- A computer data acquisition system
- 1 pressure sensor

Recommended group size:	2	Interactive demo OK?:	N

The Incredible Mass Lifter Engine

The heat engine consists of a hollow cylinder with a plunger or piston that can move along the axis of the cylinder with very little friction. The piston has a small platform attached to it for lifting masses. A short length of flexible Tygon® tubing attaches the cylinder to an air chamber (consisting of a small flask sealed with a rubber stopper) that can be placed alternately in the cold reservoir and the hot reservoir. A diagram of this mass lifter is shown in Figure 18.12.

If the temperature of the air trapped inside the syringe, cylinder, tubing, and flask is increased, then its volume will increase, causing the platform to rise. Thus, you can increase the volume of the trapped air by moving the flask from the cold to the hot reservoir. Then, when an apple has been raised through a distance *y*, it can be removed from the platform. The platform

---

This project can also be done using the PASCO scientific Heat Engine/Gas Law Apparatus (TD-8572).

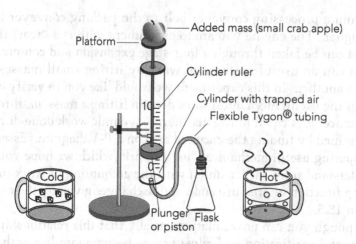

Fig. 18.12. A schematic diagram of the incredible mass lifter heat engine.

should then rise a bit more as the pressure on the cylinder of gas decreases a bit. Finally, the volume of the gas will decrease when the air chamber is returned to the cold reservoir. This causes the plunger (or piston) to descend to its original position once again. The various stages of the mass lifter cycle are shown in the following diagram.

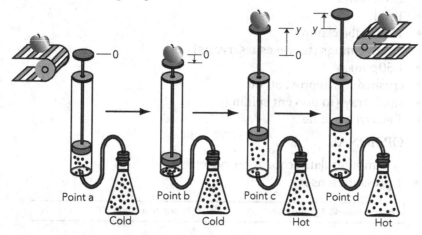

Fig. 18.13. A simplified diagram of the mass lifter heat engine at different stages of its cycle.

Before taking data on the pressure, air volume, and height of lift with the heat engine, you should set it up and run it through a few cycles to get used to its operation. A good way to start is to fill one mug with an ice and water mixture and another with preheated water at about 60–70°C. The engine cycle is much easier to describe if you begin with the plunger resting above the bottom of the cylinder. Thus, we suggest you raise the plunger to the 2 cc mark before inserting the rubber stopper firmly in the small flask. Also, air does leak out of the syringe slowly. If a large mass is being lifted, the leakage rate increases, so we suggest that you use a 50 g mass instead of an apple.

> **Warning:** If you use a larger flask or a greater temperature difference with a 10 cc syringe, the plunger might shoot out of the syringe and break!

After observing a few engine cycles, you should be able to describe each of the points *a*, *b*, *c*, and *d* of a cycle carefully, indicating which of the transitions between points are approximately adiabatic and which are isobaric.

You should reflect on your observations by answering the questions in the next activity. You can observe changes in the volume of the gas directly and you can predict how the pressure exerted on the gas by its surroundings ought to change from point to point by using the definition of pressure as force per unit area.

18.10.1. Activity: Description of the Engine Cycle

a. Predicted transition $a \mapsto b$: Close the system to outside air but leave the flask in the cold reservoir. Make sure the rubber stopper is firmly in place in the flask. What should happen to the height of the platform when you add a mass? Explain the basis of your prediction.

b. Observed transition $a \mapsto b$: What happens when you add the mass to the platform? Is this what you predicted?

c. Predicted transition $b \mapsto c$: What do you expect to happen when you place the flask in the hot reservoir?

d. Observed transition $b \mapsto c$: Place the flask in the hot reservoir and describe what happens to the platform with the added mass on it. Is this what you predicted? (This is the engine power stroke!)

e. Predicted transition $c \mapsto d$: Continue to hold the flask in the hot reservoir and predict what will happen if the added mass that is now lifted is removed from the platform and moved onto an upper conveyor belt. Explain the reasons for your prediction.

    f. Observed transition $c \mapsto d$: Remove the added mass and describe what actually happens. Is this what you predicted?

    g. Predicted transition $d \mapsto a$: What do you predict will happen if you now place the can back in the cold reservoir? Explain the reasons for your prediction.

    h. Observed transition $d \mapsto a$: Now it's time to complete the cycle by cooling the system down to its original temperature for a minute or two before placing a new mass to be lifted on it. Place the flask in the cold reservoir and describe what actually happens to the volume of the trapped air. In particular, how does the volume of the gas actually compare to the original volume of the trapped air at point a at the beginning of the cycle? Is it the same or has some of the air leaked out?

    i. Theoretically, the pressure and volume of the gas should be the same once you cool the system back to its original temperature. Why?

Determining Pressures and Volumes for a Cycle

In order to calculate the thermodynamic work done during a cycle of this engine, you will need to be able to plot a P-V diagram for the engine based on determinations of the volumes and pressures of the trapped air in the cylinder, the Tygon® tubing, and the flask at the points a, b, c, and d in the cycle.

18.10.2. Activity: Volume and Pressure Equations

    a. What is the equation for the volume of a cylinder that has an inner diameter of d and a length L?

b. Use the definition of pressure to derive the equation for the pressure on a gas being contained by a vertical piston of diameter d if the total mass on the piston including its own mass and any added mass is denoted as M? **Hints:** (1) What is the definition of pressure? (2) What is the equation needed to calculate the gravitational force on a mass, M, close to the surface of the Earth? (3) Don't forget to add in the atmospheric pressure, P^{atm}, acting on the piston and hence the gas at sea level. **Note:** The atmospheric pressure should be expressed in Pascal units in calculations.

Now that you have derived the basic equations you need, you should be able to take your engine through another cycle and make the measurements necessary for calculating both the volume and the pressure of the air and determining a P-V diagram for your heat engine. Instead of calculating the pressures, if you have the optional computer data acquisition system available, you might want to measure the pressures with a gas pressure sensor as you enter data for volume into the computer manually at key times during the cycle. You can use the experiment file L181003 for this.

18.10.3. Activity: Determining Volume and Pressure

a. Take any measurements needed to determine the volume and pressure of air in the system at all four points in the engine cycle. You should do this rapidly to avoid air leakages around the piston and summarize the measurements with units in the space below.

b. Next you can use your measurements to calculate the pressure and volume of the system at point a. Show your equations and calculations in the space below and summarize your results with units. Don't forget to take the volume of the air in the Tygon® tubing and can into account! **Hint:** You can use the electronic balance to measure the mass of the plunger.

$P_a =$ _____ $V_a =$ _____

c. Use the measurements at point b to calculate the total volume and pressure of the air in the system at that point in the cycle. Show your equations and calculations in the space below and summarize your results with units.

$P_b = $ _____ $V_b = $ _____

d. What is the height, y, through which the added mass is lifted in the transition from b to c?

e. Use the measurements at point c to calculate the total volume and pressure of the air in the system at that point in the cycle. Show your equations and calculations in the following space and summarize your results with units.

$P_c = $ _____ $V_c = $ _____

f. Remove the added mass and make any measurements needed to calculate the volume and pressure of air in the system at point d in the cycle. Show your equations and calculations in the space below and summarize your results with units.

$P_d = $ _____ $V_d = $ _____

g. We suspect that transitions from a↦b and from c↦d are approximately adiabatic. Explain why.

h. You should have found that the transitions from $b \mapsto c$ and from $d \mapsto a$ are isobaric. Explain why this is the case.

Finding Thermodynamic Work from the P-V Diagram

In the next activity you should draw a P-V diagram for your cycle and determine the thermodynamic work for your engine.

18.10.4. Activity: Plotting and Interpreting a *P-V* Diagram

a. Fill in the appropriate numbers on the scale on the graph frame that follows and plot the *P-V* diagram for your engine cycle. Alternatively, generate your own graph using a computer graphing routine and affix the result in the space below.

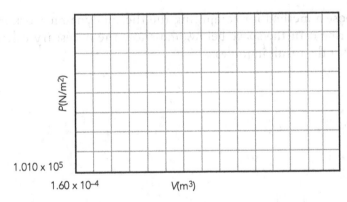

b. On the graph in part a. label each of the points on the cycle (*a, b, c,* and *d*). Indicate on the graph which of the transitions (*a↦b, b↦c,* etc.) are adiabatic and which are isobaric.

Next you need to find a way to determine the area enclosed by the *P-V* diagram. The enclosed area doesn't change very much if you assume that *P* is approximately a linear function of *V* for the adiabatic transitions. By making this approximation, the figure is almost a parallelogram so you can obtain the enclosed area using one of several methods. Three of the many possibilities are listed on the next page. *Creative students have come up with even better methods than these, so you should think about your method of analysis carefully.*

Method I

Since the pressure doesn't change from point *b* to point *c* you can take the pressure of those two points as a constant pressure between points. The same holds for the transition from *d* to *a*. This gives you a figure that is approximately a parallelogram with two sets of parallel sides. You can look up and properly apply the appropriate equation to determine the net thermodynamic work performed.

Method II

Display your graph with a grid and count the boxes in the area enclosed by the lines connecting points *a, b, c,* and *d*. Then multiply by the number of joules each box represents. You will need to make careful estimates of fractions of a box when a "leg" of a cycle cuts through a box.

Method III

Fit a straight line to each of the starting and ending points for the four transitions in the cycle. Each equation will give you a function relating *P* and *V*. Perform an integral for each of these equations since

$$\oint P dV = \int_a^b P dV + \int_b^c P dV + \int_c^d P dV + \int_d^a P dV$$

18.10.5. Activity: Comparing the Thermodynamic and Useful Mechanical Work

a. Choose a method for computing the thermodynamic work in joules, describe it in the space below, and show the necessary calculations. Report the result in joules.

b. What is the equation you need to use to calculate the useful mechanical work done in lifting the mass from one level to another?

    c. Use the result for the height that the mass is lifted in the power stroke of the engine to calculate the useful mechanical work performed by the heat engine.

    d. How does the thermodynamic work compare to the useful mechanical work? Please use the correct number of significant figures in your comparison (as you have been doing all along, right?)

The Incredible Mass Lifter Engine Is Not So Simple

Understanding the stages of the engine cycle on a P-V diagram is reasonably straightforward. However, it is difficult to use equations for adiabatic expansion and compression and the ideal gas law to determine the temperature (and hence the internal energy) of the air throughout the cycle. There are reasons for this. First, there is some friction between the syringe walls and the plunger which can impede the plunger's movement. Second, the mass lifter engine is not well insulated and so the air that is warmed in the hot reservoir transfers thermal energy through the cylinder walls. Thus, the air in the flask and in the cylinder are probably not at the same temperature. Third, air might leak out around the plunger. This means that the number of moles of air could decrease over time. You can observe this by noting that in the transition from point *d* to point *a* the piston can actually end up in a lower position than it had at the beginning of the previous cycle.

    A real working heat engine differs from our mass lifter because it has built in linkages that use the action of the engine to alternately place the working medium in contact with a hot and then a cold reservoir automatically.

    In the next session you will conduct a theoretical investigation of the stages of operation of an ideal heat engine that is much more efficient than the mass lifter.

CARNOT AND STIRLING ENGINE CYCLES

18.11. AN IDEAL HEAT ENGINE

So far in this unit you discovered that all heat engines must operate in a cycle, with the system returning to its starting state so that it can repeatedly do work. We also found that every heat engine that we investigated only converted part of the thermal energy transferred to it from a high temperature reservoir into work. The rest, the "waste" thermal energy, had to be transferred to a low-temperature reservoir.

    Is it possible to devise very efficient engines that transfer little or perhaps no "waste" thermal energy to a cold reservoir? Such an engine would be extremely useful, even during an "oil glut." If we were able to use more efficient turbine engines for electrical power generation, less fuel would be burning and less carbon dioxide would be released to the atmosphere. This would slow down the rate of global warming. The design of more efficient engines is of great interest regardless of the price and availability of oil.

Before the development of the science of thermodynamics, a brilliant French engineer named Sadi Carnot (1796–1832) proposed the most efficient possible engine cycle. In Carnot's "ideal" engine cycle a monatomic ideal gas consisting of point particles undergoes two adiabatic and two isothermal processes. This Carnot engine cycle will be described in more detail in Section 18.15. However, in order to understand the details of the Carnot engine cycle, you will need to learn more about molar heat capacities of gases held at constant pressure and at constant volume. In particular, you need to learn more about how these heat capacities influence adiabatic processes in which no thermal energy is transferred into or out of a medium.

18.12. MOLAR HEAT CAPACITY FOR AN IDEAL GAS

In studying how gases behave when they give up or gain thermal energy in an engine cycle involving adiabatic processes, it is helpful to define *molar heat capacity* as a measure of the amount of thermal energy a mole of gas absorbs when it undergoes a temperature change. Although you will use ideal gases in your calculations, the concept of heat capacity is easily generalized to the case of non–ideal gases and will also allow you to learn more about the workings of real heat engines using real gases.

In general, the heat capacity for a solid or liquid is defined by the expression

$$C \equiv \frac{Q}{\Delta T} \tag{18.6}$$

where Q is the thermal energy transferred to the material and ΔT is the temperature change it undergoes as a result of that transfer of thermal energy. If a sample of a gas contains n moles, the *molar heat capacity* of the gas can be defined as

$$C_{\mathrm{molar}} \equiv \frac{1}{n}\left(\frac{Q}{\Delta T}\right) \tag{18.7}$$

Note: We use two definitions having to do with heat capacity. Do not confuse *molar heat capacity*, which we have just defined that is represented by an uppercase "C," with *specific heat capacity*, *c*, that we used in Unit 16 to refer to thermal energy absorption per unit mass rather than per mole.

If thermal energy is transferred to a gas, the gas can either undergo a significant volume change (if left at a constant pressure) or a significant pressure change (if confined to a constant volume). Measurements of molar heat capacity in a gas yield different values at constant volume than at constant pressure. We must keep this in mind if we want to relate the molar heat capacity to the change in the internal energy of the gas as a result of absorbing thermal energy.

First, let's consider the constant volume case. This case, for an ideal gas, is one of the few circumstances in which the change in internal energy of the gas is the same as the energy transferred to it. This can be seen from the first law of thermodynamics since

$$\Delta E^{int} = Q - W = Q - P\Delta V = Q \text{ (since } \Delta V = 0)$$

For an ideal gas we can write the internal energy as

$$E^{int} = \frac{3}{2} NkT = \frac{3}{2} nRT \qquad (18.8)$$

where N is the number of molecules, $n = N/N_A$ is the number of moles of the gas, and N_A is Avogadro's number. Because the volume of our gas remains constant,

$$\Delta E^{int} = Q = C_V n\Delta T$$

or

$$C_V = \frac{1}{n}\left(\frac{\Delta E^{int}}{\Delta T}\right) = \frac{3}{2} R \qquad (18.9)$$

where C_V is the *molar heat capacity at constant volume* and R is the universal gas constant. We can define C_V for a non-ideal diatomic gas using the same approach, but it will not have the value $3/2\,R$.

In order to understand the Carnot cycle as an ideal heat engine cycle, we must explore the nature of adiabatic expansions and the work associated with them. Adiabatic expansions are a function of the ratio of the molar heat capacity at constant volume and that at constant pressure. It can be shown mathematically that the relationship between these two heat capacities is

$$C_P = C_V + R \qquad (18.10)$$

If this equation is valid, then obviously C_P is greater than C_V. Another way of saying this is that when a given amount of thermal energy is transferred to a gas, the temperature of the gas will rise more when the volume is held constant than when the pressure is held constant. How come? This relationship can be explained using kinetic theory.

For simplicity, let's consider a mole of ideal gas that has thermal energy transferred to it. If the volume is held constant, the gas does no work; the thermal energy is absorbed so that all of the thermal energy goes into speeding up the molecules. Since the temperature is directly related to the average speed of the gas molecules, all the added heat or thermal energy goes to raising the temperature of the gas. This is *not* the case for the situation when the pressure of the gas is held constant. Some of the transferred thermal energy is used up in allowing the gas to expand and hence do work on its surroundings. Less energy is left over to speed up the molecules. Hence at constant pressure, the temperature rise is less than it is at constant volume. Thus, C_P is greater than C_V.

18.13. ADIABATIC CHANGES AND THE P-V DIAGRAM

To understand the ideal heat engine proposed by Carnot, we will calculate the work done when a monatomic ideal gas expands or is compressed adiabatically so that no thermal energy is transferred to or from the gas. In general, as a gas expands to a new volume and does work, the pressure is not constant.

Thus, it is not possible to evaluate the work integral $\int P dV$ unless we know how the pressure varies with volume. It is precisely this relationship we now seek. In Activity 18.4.2, you showed that for an ideal monatomic gas consisting of point particles, the relationship between temperature and volume is

$$T_2^{3/2} V_2 = T_1^{3/2} V_1 \tag{18.11}$$

for an adiabatic process. It is a simple matter to use the ideal gas law to show that the relationship between pressure P and volume V for an adiabatic expansion or compression is given by

$$P_2 V_2^{\gamma} = P_1 V_1^{\gamma} \tag{18.12}$$

where the exponent $\gamma = 5/3$.

However, it is not so obvious that the exponent of 5/3 turns out to be the ratio of the heat capacities, C_P/C_V for an ideal monatomic gas. In the next activity you will derive these relationships for adiabatic expansions in order to use them to analyze the work done by our "perfect" engine.

The relationship between P and V for an adiabatic expansion of a monatomic ideal gas can be derived by using the ideal gas law, some definitions, and the first law of thermodynamics. To complete the derivation you will need to use the relationship between the heat capacities of an ideal gas at constant pressures and constant volumes. In Section 18.12 we explained why the heat capacity of an ideal gas is greater at constant pressure than at constant volume. We then stated without formal proof that C_P can be shown to be greater than C_V by an amount equal to the universal gas constant R. You will also have to use this relationship in your derivation.

18.13.1. Activity: Does P-V^{γ} = Constant for an Adiabatic Expansion?

a. For an ideal gas described by $PV = nRT$, use the fact that for small changes in pressure and volume $\Delta(PV) \approx P \, \Delta V + V \, \Delta P$ and the relationship $C_P - C_V = R$ to show that

$$n\Delta T \approx \frac{P\Delta V + V\Delta P}{R} = \frac{P\Delta V + V\Delta P}{C_P - C_V}$$

b. For an adiabatic expansion, $Q = 0$, so the first law reduces to $\Delta E^{\text{int}} = Q - P\Delta V = -P\Delta V$. Using the fact that for any change $\Delta E^{\text{int}} = C_V n\Delta T$ for a gas of noninteracting particles, show that

$$n\Delta T = -\frac{P}{C_V} \Delta V$$

c. Combine your results for a. and b. to show that

$$\frac{\Delta P}{P} + \frac{C_P}{C_V} \frac{\Delta V}{V} = 0$$

d. Show that in the limit of very small changes of variables,

$$\frac{dP}{P} + \frac{C_P}{C_V} \frac{dV}{V} = 0$$

can be integrated from an initial P_1 and V_1 to a final P_2 and V_2 to yield

$$\ln(P_1) + \frac{C_P}{C_V} \ln(V_1) = \ln(P_2) + \frac{C_P}{C_V} \ln(V_2)$$

e. Show that this confirms our suspicion that $P_2 V_2{}^\gamma = P_1 V_1{}^\gamma$ with $\gamma = (C_P / C_V) = 5/3$.

f. Use the result in part e. in conjunction with the ideal gas law to show that $T_2 V_2{}^{\gamma-1} = T_1 V_1{}^{\gamma-1}$.

18.14. WORK IN ADIABATIC AND ISOTHERMAL EXPANSIONS

We are still aiming at determining the work done in each cycle of the ideal Carnot engine, which we are going to analyze soon. Thus, we need to consider the work associated with both adiabatic and isothermal expansions.

Adiabatic Work

Consider the work done in the adiabatic expansion of a mole of a monatomic gas. Assume that $\gamma = 5/3$ for the gas.

18.14.1. Activity: Work in an Adiabatic Expansion

a. The result you just obtained in Section 18.13 can be written $PV^\gamma = P_1 V_1^\gamma$ or $P = (V^{-\gamma})P_1 V_1^\gamma$ for *any* point in an adiabatic expansion. Use this to show that this adiabatic P-V relationship in conjunction with the equation for work, $W = \int P dV$ yields an adiabatic work equation of

$$W^{\text{adiabatic}} = \frac{(P_1 V_1^\gamma)(V_2^{1-\gamma} - V_1^{1-\gamma})}{(1-\gamma)}$$

b. Calculate the work done when one mole of 300 K gas expands adiabatically from an initial pressure of $8.31 \times 10^2 \, \text{N/m}^2$ and volume of $3.00 \, \text{m}^3$ to a final pressure of $3.02 \times 10^2 \, \text{N/m}^2$, a volume of $5.51 \, \text{m}^3$.

Isothermal Work

Consider the work done in the isothermal expansion of n moles of a monatomic gas.

18.14.2. Activity: Work in an Isothermal Expansion

a. Use the fact that $W = \int P dV$ in conjunction with the ideal gas law to show that for an isothermal expansion

$$W^{\text{isothermal}} = nRT \int_{V_1}^{V_2} \frac{1}{V} \, dV = nRT \ln\left(\frac{V_2}{V_1}\right)$$

Hints: (1) Consider why it is legitimate to pull the n, R, and T terms out of the integral for any isothermal expansion or compression. (2) What is the expression for the integral of dV/V that you derived in Activity 18.4.1d?

b. Calculate the work done when one mole of a 300 K gas expands isothermally from an initial pressure of 2.49×10^3 N/m^2 and volume of 1.00 m^3 to a final pressure of 8.31×10^2 N/m^2 and volume of 3.00 m^3.

18.15. THE CARNOT ENGINE CYCLE

Let us return to a consideration of the Carnot cycle, which can be shown to be the most efficient possible heat engine cycle. It consists of four elements pictured below on a P-V diagram: (1) work done by the gas in an isothermal expansion from $A \mapsto B$ in a piston at T_H; (2) work done by the gas in an adiabatic expansion from $B \mapsto C$ in which the gas is allowed to cool to T_C; (3) work done on the gas in an isothermal compression of the gas from $C \mapsto D$ at T_{cold}; and (4) work done on the gas in an adiabatic compression of the gas from $D \mapsto A$ while temperature rises back to T_H.

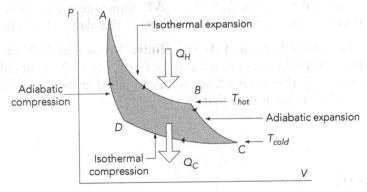

Fig. 18.14. A Carnot cycle consisting of two adiabatic and two isothermal processes.

A Sample Carnot Cycle

Here is a specific example of a Carnot cycle involving 1.00 moles of an ideal monatomic gas for which $\gamma = 5/3$. It has four "legs." You will be using this sample cycle data in Activity 18.15.1 to make a series of specific calculations that should help you understand the relationship between the thermal energy transfers and the temperatures of the reservoirs for a Carnot engine.

Isothermal Expansion $A \mapsto B$

Point A: The gas is confined to a volume of 1.00 m³ and a pressure of 2.49×10^3 N/m². It is initially at equilibrium with a heat reservoir at a temperature of 300 K. (A heat reservoir is a source of energy that is recharged so it stays at the same temperature no matter how much thermal energy is transferred from it.)

Point B: The gas is allowed to do work on its surroundings by expanding isothermally to a new volume of 3.00 m³ and a pressure of 8.31×10^2 N/m².

Adiabatic Expansion $B \mapsto C$

Point C: The gas is thermally isolated by wrapping the piston in an insulating material and is allowed to do more work and expand further adiabatically until it has cooled to a temperature of 200 K. In this adiabatic process the pressure drops to 3.02×10^2 N/m² and the volume increases to 5.51 m³.

Isothermal Compression $C \mapsto D$

Point D: The gas is placed in thermal contact with a heat reservoir at 200 K and work is done to compress it isothermally to a volume of 1.84 cubic meters at an increased pressure of 9.05×10^2 N/m².

Adiabatic Compression $D \mapsto A$

Point A: Again: The gas is isolated thermally by insulating it. Then work is done on it to compress it until it reaches a temperature of 300 K and a volume of 1.00 m³ once again.

18.15.1. Activity: Carnot Cycle Analysis

a. Calculate the ΔE^{int}, Q, and W values for each of the parts of the sample Carnot cycle. Make use of the First Law ($\Delta E^{int} = Q - W$) when you can and recall that $\Delta E^{int} = nC_V \Delta T$. Show the equations and calculations and then summarize the results in the blanks that follow:

1. Isothermal Expansion $A \mapsto B$. **Hints:** Recall that ΔE^{int} can be calculated from the temperature change from A to B. You should be able to use the isothermal work equation and calculations you did in Activity 18.14.2 to determine that $W_{AB} = 2739$ J.

$A \mapsto B$:

$\Delta E^{int} = $ _____ $Q = $ _____ $W = $ _____

2. Adiabatic Expansion $B \mapsto C$. **Hint:** You can use the fact that no thermal energy is transferred to the engine so $Q = 0$ J and that ΔE^{int} can be calculated from the known temperature change between points B and C.

$B \mapsto C$:

$\Delta E^{int} = $ _____ J $Q = $ _____ J $W = $ _____ J

3. Isothermal Compression $C \mapsto D$

$C \mapsto D$:

$\Delta E^{int} = $ _____ J $Q = $ _____ J $W = $ _____ J

4. Adiabatic Compression $D \mapsto A$

$D \mapsto A$:

$\Delta E^{int} = $ _____ J $Q = $ _____ J $W = $ _____ J

b. Show that the efficiency of this Carnot cycle is $\eta = 0.33$. Write the equation that defines heat engine efficiency and also show your calculations.

c. Compare the quantities listed below:

$|Q_H| = $ _____ J $T_H = $ _____ $\dfrac{|Q_H|}{T_H} = $ _____

$|Q_C| = $ _____ J $T_C = $ _____ $\dfrac{|Q_C|}{T_C} = $ _____

d. Do you see any relationships between the thermal energy transfers and the temperatures tabulated above? Explain.

e. Can you rewrite the efficiency of your Carnot cycle in terms of the temperature of the two reservoirs?

18.16. THE CARNOT EFFICIENCY

From your investigation of the Carnot cycle you should have discovered that

$$\frac{|Q_H|}{|Q_C|} = \frac{T_H}{T_C} \quad \text{so that} \quad \frac{|Q_H|}{T_H} = \frac{|Q_C|}{T_C}$$

Carnot recognized that this meant that efficiency, η ("eta"), of his ideal cycle could be described by the equations

$$\eta_{Carnot} = \frac{|W|}{|Q_H|} = \frac{|Q_H| - |Q_C|}{|Q_H|} = 1 - \frac{|Q_C|}{|Q_H|} = 1 - \frac{T_C}{T_H}. \quad (18.13)$$

Thus, the efficiency of a Carnot engine *depends only on the temperature ratio* between the hot and the cold reservoir. The bigger the ratio, the more efficient the engine. This increase in efficiency with increasing temperature differences holds true for other heat engine cycles, but no cycle has ever been found that is *more* efficient than the Carnot cycle for a given T_C/T_H. What is the secret behind the Carnot cycle's efficiency? In order to answer this question scientists have introduced a new concept called entropy and studied how it changes during various engine cycles. Unfortunately, we do not have time to develop this concept.

18.17. THE STIRLING ENGINE AND THE SECOND LAW OF THERMODYNAMICS

The Stirling Engine

Carnot began working on engines in hopes of improving the efficiency of the steam engine. Although his concept of the ideal heat engine was a rare achievement that laid the groundwork for the first and second laws of thermodynamics, the internal combustion engine used in the cars we drive is far from ideal in its efficiency. The Stirling Engine Cycle proposed by the Reverend Robert Stirling of the Church of Scotland in 1816 is considerably closer in its design to the Carnot engine. In a Stirling engine a piston linked to a displacement system shuffles gas back and forth between hot and cold reservoirs. The expansion and contraction of the gas as it is heated and cooled drives the engine. In the Stirling engine waste heat (thermal energy) transferred back to the engine's surroundings is recycled in an ingenious way that improves efficiency. A modern working model of the early Stirling engine enables us to explore the quantitative behavior of a Carnot-like engine.

For observing the operation of the Stirling engine you will need:

EITHER:

- 1 miniature Stirling engine
- 6 oz. of denatured alcohol
- 2 ice cubes

AND/OR

- 1 Visible Stirling Engine
- 1 ceramic coffee mug
- hot water

Recommended group size:		All	Interactive demo OK?:		Y

18.17.1. Activity: Stirling Engine Efficiency

a. Follow the instructions that come with the Stirling engines and operate it. Examine the engine and try to explain the elements of a basic cycle of the engine. Where is the hot reservoir? The cold reservoir?

b. Assuming that the equation describing the efficiency of the engine is approximately the same as that for the Carnot engine so that

$$\eta_{Carnot} = 1 - \frac{T_C}{T_H}$$

What do you predict will happen to the engine if an ice cube is placed in contact with the cold reservoir?

c. Place the ice cube in contact with the cold reservoir and describe what actually happens to the operation of the engine.

The Second Law of Thermodynamics

Time does not permit you to gain direct experience with the measurement of the efficiencies of the many real heat engines that have been devised. However, experience with hundreds of types of heat engines led scientists in the nineteenth century to conclude the following:

> It is impossible to construct a heat engine that absorbs energy via thermal processes from a high temperature reservoir and delivers work with 100% efficiency.

Furthermore, theoretical considerations drawn from experience with heat engines have led to the additional conclusion that

> It is impossible to construct a heat engine that is more efficient than a Carnot engine operating between the same two temperatures.

These are two of many different statements of the *Second Law of Thermodynamics*. There are several other statements that require familiarity with the concept of entropy to understand. We hope you will learn more about this concept in the future.

UNIT 28: RADIOACTIVITY AND RADON

Unstable atomic nuclei give off alpha, beta, or gamma particles. One characteristic of all three of these decay products is that they can create ions by removing electrons from atoms or molecules they encounter. The cloud chamber in the photograph contains a mixture of air and vapor (often alcohol) along with a collection of radioactive nuclei that decay by emitting alpha particles. As these alpha particles ionize air molecules along their path, the vapor in the chamber condenses on the ions, creating a cloud trail much like the contrails seen behind high flying airplanes. A picture of some of the alpha trails is shown in the inset. In this unit you will learn about the characteristics of alpha, beta, and gamma particles. You will also learn about some changes that decaying nuclei undergo and about the health effects of ion-izing radiation. You will use a common Geiger counter to detect radioactive radon gas that is considered a natural radiation health hazard in certain parts of the United States.

UNIT 28: RADIOACTIVITY AND RADON

Motionless in appearance, matter contains births, collisions, murders and suicides. It contained dramas subjected to implacable fatality; it contained life and death. Such were the facts which the discovery of radioactivity revealed. Philosophers had only to begin their philosophy all over again and physicists their physics.   Eve Curie (1937)

OBJECTIVES

1. To learn about naturally occurring ionizing radiation and measure its level of activity at different locations.

2. To develop a mathematical model to describe radioactive decay and to understand the physical meaning of the decay constant and the half-life for a given decay process.

3. To use the mathematical model of decay processes to predict counting rate changes over time for very long-lived radioactive nuclei and for relatively short-lived radioactive nuclei.

4. To observe the statistical fluctuation of counts from a naturally occurring radioactive material with a very long half-life.

5. To understand more about radiation safety in terms of the relationship between the properties of each type of radioactive material and its potential to do biological damage, and to learn why radon and its daughter elements pose a health risk.

6. To use the mathematics of radioactive decay processes to construct a theoretical prediction of how the number of radon 222 daughters increase and then decrease in amount over time and compare that prediction to actual data.

28.1. OVERVIEW

The term radiation is commonly used to describe invisible forms of energy moving through space. A familiar form of radiation is the sound wave, which can move invisibly from place to place. Another common form of radiation is the electromagnetic wave. These waves include low energy radio waves, microwaves, visible light, X-rays, and high energy gamma rays. Electromagnetic waves appear to be made up of oscillating electric and magnetic fields like those you have been studying in the last few units. Other forms of radiation include the tiny alpha particles and beta particles that are ejected from the nuclei of certain elements that are called radioactive if they emit radiation spontaneously.

In this unit we are going to investigate ionizing radiation. Ionizing radiation is defined as any type of radiation that is energetic enough to knock electrons out of atomic orbits around the nuclei of atoms. An atom with a missing electron is said to be "ionized".

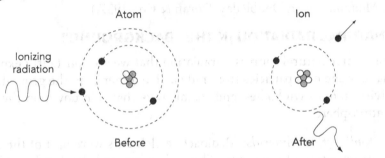

Fig. 28.1. Schematic of ionizing radiation in the form of a gamma ray that knocks an electron out of an atom, leaving the atom ionized.

Ionizing radiation passes through our bodies all the time. It comes from cosmic rays from other parts of the universe and from radioactive atomic nuclei that exist naturally in the materials in our surroundings and in our own bodies. For example, radioactive radon gas is in the air we breathe. In addition to naturally occurring radiation, we are exposed to man-made sources of ionizing radiation, including fallout from the testing of nuclear weapons, releases from nuclear power plants, and radioactive materials in tobacco smoke. Ionizing radiation is of special interest to us because exposure to it can do damage.

In this unit you will measure the relative level of the background radiation in and around your classroom and explore ways to describe radioactive decay rates mathematically. Then you will learn about some of the properties of atomic nuclei and about several common types of radioactive decay. You will study a chain of successive radioactive decays in which a parent nucleus, uranium 238, emits ionizing radiation and is thereby transformed into a *daughter* nucleus that in turn decays into another daughter nucleus, and so on. It is this decay series that leads to the creation of a form of radon gas called radon 222. Finally, you will learn more about radon, why it is a health problem, and how to measure its presence by collecting and counting radiation from two radon daughter elements. You will also learn to calculate the expected rate of creation and decay of these radon daughter elements over time.

RADIATION AND RADIOACTIVE DECAY

28.2. THE DISCOVERY OF RADIOACTIVITY

Just before the turn of the century, Henri Becquerel examined the salts of an unusually heavy rare metal called uranium and found that they caused photographic film surrounded by black paper to darken. He also found that the uranium salts caused an electroscope to discharge. The uranium salts gave off strange, tiny particles of radiation, or rays. Marie Curie named this process of radiation emission *radioactivity*. Marie and her husband, Pierre, discovered that most of the radiation in the uranium salts came from a new element they called radium which had an atomic mass of 226. Over a span of four years ending in 1902 the Curies labored in a dingy shed with a leaky roof to extract one tiny gram of radium from eight tons of uranium ore residue. The biography of Madame Curie, written by her daughter, Eve, contains a fascinating account of the discovery of radium. It makes good summer reading! (Eve Curie, *Madame Curie*, Doubleday, Doran & Co., 1937.)

28.3. NATURAL RADIATION IN THE "BACKGROUND"

There are three natural sources of radiation that we are constantly exposed to: gamma rays and beta particles from radioactive minerals in the ground, the radioactivity in our own bodies, and cosmic rays that rain down on the Earth's upper atmosphere.

1. *Radioactive Minerals*: Radioactive elements were part of the original composition of the earth. These elements and the daughter elements formed as a result of their decay emit ionizing gamma and beta radiation that cause most of the exposure of humans to natural radiation. The major sources of naturally radioactive elements are potassium, thorium, and uranium.

2. *Radioactive Materials in the Human Body*: The radioactive elements found in the human body come mostly from the ingestion of food, water, and tobacco smoke that contain them. Potassium and radium (and its decay products) are the most common radioactive elements that are ingested. Some additional elements come from the inhalation of radon, an airborne noble gas, and daughter products that become attached to dust particles. The alpha particles emitted by internal sources of radiation are the source of most of the exposure to ionizing radiation from materials in the body.

3. *Cosmic Rays*: About 2×10^{18} primary cosmic ray particles, consisting mostly of protons having energies of more than a billion electron volts, are incident on the earth's atmosphere each second. Most of the primary cosmic rays interact with atoms in the atmosphere and produce hundreds of secondary radiation particles, such as muons, electrons, and gamma rays. Sometimes these secondary particles arrive at the Earth's surface in bursts or showers.

28.3.1. Activity: Predicting Background Radiation

Suppose you were to take a radiation monitor outside and count the number of particles of ionizing radiation detected by the Geiger tube. Do you think you'll detect more radiation or less radiation than you would if you count inside with the Geiger tube? On the basis of the description of natural radiation you just read, explain your prediction.

28.4. MEASURING BACKGROUND RADIATION

Let's use a small hand-held radiation monitor to measure the relative number of background counts in various places indoors and outdoors near your classroom. This will also give you a chance to reconsider the effects of statistical fluctuations on counting rates, which you observed in the second unit. For this activity you will need the following items:

- 1 battery-operated radiation monitor w/ digital readout
- 1 digital stopwatch

OPTIONAL

- 1 computer data acquisition system
- 1 radiation software

Recommended group size:	4	Interactive demo OK?:	N

The radiation monitor contains a small Geiger tube that is capable of detecting gamma and beta particles that come either from secondary particles produced by cosmic rays or from the decay of radioactive elements in nearby radioactive materials. You'll be learning more about how the Geiger tube works in the next several activities.

You are to monitor background radiation both inside and outside.

Outside Background Radiation

Let's start with outside measurements. To monitor radiation outside, go outdoors and turn the radiation monitor on. A click and/or light flash will occur whenever a gamma ray or beta particle passes through the Geiger tube in the radiation monitor and ionizes atoms in the gas inside the tube. Use the stopwatch to record counts per minute for 1 minute. You can read a digital display or count audio clicks or indicator light flashes for a minute.

28.4.1. Activity: Background Radiation Outside

Determine the number of counts/minute for four trials each lasting one minute. Record the results below and find the average and a standard deviation for your data.

Trial #	Counts/Min
1	
2	
3	
4	
Average	
Standard deviation	

Background Radiation in the Classroom

In order to monitor background radiation in the classroom you should set up your computer-based Radiation Counting System as shown in Figure 28.2. When appropriate event-counting software is loaded into the computer, you can take and display data automatically.

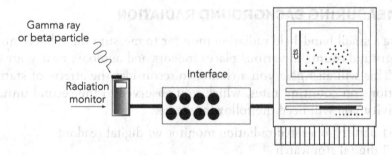

Fig. 28.2. A computer-based radiation counting system.

You should consult the manuals for your computer-based laboratory system and the associated software for detailed instructions for setting up the system and using the software.

The background counts may vary from place to place in the classroom. You will need to determine the background counting rate at your own table in order to correct future data. **Note:** When you record more data in the future you should attempt to locate your radiation sensor in the same place.

28.4.2. Activity: Background Radiation at Your Table

a. Place the Radiation Monitor in a convenient place on your lab table. Record the number of counts/minute for 10 one-minute long trials. Record the results below.

Trial	cts/min	Trial	cts/min
1		7	
2		8	
3		9	
4		10	
5		Average	
6		Standard deviation	

b. In Unit 2 you explored the relationship between the standard deviation, σ, and the square root of the average number of counts in a counting interval. You should have found them to be about the same. Does this rule of thumb hold here? Compare the square root of the average number of counts/min to the calculated standard deviation for your data set. Comment on the result of the comparison.

    c. Does there appear to be a statistically significant difference between the average of the two sets of counts recorded inside and outside? If so, was the difference what you predicted? Explain.

28.5. RADIOACTIVE DECAY RATES

Radioactivity is now understood as a phenomenon in which neutrons and protons in a nucleus lose potential energy. Every once in a while, a nucleus in a collection of radioactive atoms ejects a tiny wave-like particle spontaneously. Typically, the particle is either a gamma ray, a beta particle, or an alpha particle. After the radiation is emitted, the neutrons and protons that are attracted to each other by the strong nuclear force move closer together. This process is similar to the one in which a mass loses potential energy by falling closer to the center of the earth.

Radioactivity is a statistical process in which each undecayed nucleus is in an unstable state. A series of slight disturbances can lead to a decay but not at a definite time. The best we can say is that a radioactive nucleus has a certain probability of undergoing a decay during a given period of time Δt. Let's predict what will happen to a large collection of radioactive nuclei *each of which has a known probability of decay*. We can do this by considering the behavior of a large collection of dice that are rolled once each minute.

For this activity, you will need:

- 300 dice

28.5.1. Activity: Predicting and Measuring the Decay of Dice

    a. Assume that a *single* die represents a radioactive nucleus that hasn't undergone a decay yet. Suppose it is rolled sometime during a one-minute period. What is the relative probability that a "2" will come up during the first minute? **Note:** In probability theory relative probabilities are always numbers between 0 and 1 with 0 indicating that an event never happens and 1 indicating that an event is certain to happen.

        Predicted relative
        probabilty of rolling a "2" = _____

    b. Suppose you have a collection of $N(0) = 300$ undecayed dice at time 0. In general, the symbol $N(t)$ represents the number of dice at some later time, t. If the dice in the collection are rolled once during the first minute, how many dice, ΔN, will have a "2" appear face up and hence will have decayed during the first minute?

        Predicted number of dice that
        decay in the first roll (out of 300) = _____

c. How many dice, $N(1)$, do you predict are left at the end of the first minute (the first roll) and what is the rate of decay $\Delta N/\Delta t$ in {dice/min} during the first minute?

Predicted number of
undecayed dice after one minute = _____ [dice/min]

d. If you remove the "decayed" dice (that is, the "2's") and roll the remaining dice once again during the second minute, how many dice do you predict will decay during the second minute? How many dice $N(2)$, do you predict will be left at the end of the second minute? What do you predict the rate of decay, $\Delta N/\Delta t$, during the second minute is?

e. Create a spreadsheet to calculate the predicted number of undecayed dice, $N(t)$ and the decay rate $\Delta N/\Delta t$ for the first 15 minutes, using the format suggested in Figure 28.3. Note that columns D and E are included to record observed results. *Be sure to format your columns as shown in the sample and save this file. You will be using it for other calculations.*

	A	B	C	D	E
2	THEORETICAL PREDICTIONS			OBSERVATIONS	
3					
4	t [min]	Predicted Undecayed Dice N(t) [dice]	Predicted Decayed Dice ΔN(t) [dice/min]	Observed Undecayed Dice N(t) [dice]	Observed Decayed Dice ΔN(t) [dice/min]
5	0	300	50		
6	1	250	42		
7	2				
8	3				
9	4				
10	5				
11	6				
12	7				
13	8				
14	9				
15	10				
16	11				

The number of undecayed dice left is B5-C5. This equation can be copied down.

Since 1/6th of the dice decay each minute, theoretically, the number of dice that decay in the first minute (the value of ΔN(t) contained in cell C5) should be calculated as B5/6.

Fig. 28.3. Affix your actual spreadsheet over this sample.

f. Let's see how good your theoretical predictions are. You and your classmates should roll the 300 dice during the first minute. Remove all the dice that have a "2" on the top face representing decayed nuclei. How many are left? Repeat the procedure with the remaining dice 14 more times and fill the results into column (E) of the spreadsheet both in the printout above and in your computer version of it.

g. Do the results match the theoretically expected results exactly? If not, attempt to explain why there are differences.

Exponential Dice Decay

You should have predicted already that the number of dice expected to decay during each roll is proportional to the number of undecayed dice in the collection. This type of proportionality is the same mathematically as that which describes capacitor decay (Unit 24) and cooling (Unit 16). In both capacitor decay and cooling a negative exponential function provides a good analytical mathematical model for the process. Is an exponential function of the form $N(t) = N(0) e^{-\lambda t}$ where λ is a decay constant a good model for dice decay?

Once you find a value of the decay constant λ that describes your dice decay, you can also determine the half-life, $T_{1/2}$, of the dice collection. Recall that half-life is defined as the time it takes for something to decay to half of its original amount.

28.5.2. Activity: Graphing Decay & Determining Half-Life

a. Create a new spreadsheet for your mathematical model. The first two columns should contain the time values (1 s to 15 s) and the corresponding number of undecayed dice as a function of time that you observed in part f. of Activity 28.5.1. In the third column enter the exponential decay equation.

b. Examine the graph of $N(t)$ vs. t. What is the approximate time it takes for the number of dice to decay from the original number of 300 to 150? How much additional time does it take for the number of dice to decay from 150 dice to 75 dice? What additional time do you think it will take for the dice to decay from 75 dice to 38 dice? What is the half-life of the dice if they are rolled once a minute?

c. Suppose the dice decay constant were doubled so the probability of decay in each minute was twice as great. (This could be simulated by claiming that upon each roll of the dice, the decay occurs whenever a "2" or a "6" comes up.) What would happen to the half-life of a collection of dice rolled once a minute? In general, how do you think the decay constant, λ, is related to half-life, $T_{1/2}$? Do you expect it to be directly proportional, inversely proportional, or what?

28.6. THE MATHEMATICS OF RADIOACTIVE DECAY

Those of you with a flair for mathematics will be delighted to discover that we can derive the mathematical equations describing radioactive decay using integral calculus. These equations can be used to do some important calculations in preparation for your observation of the counting rates associated with the decay of radon and its daughters.

Your experience with the decay of dice should convince you that the number of radioactive nuclei expected to decay in a given time is proportional to the number of undecayed nuclei at a given time, $N(t)$, and to the amount of time over which the decay is measured. The constant of proportionality is represented by the decay constant λ. Thus for a very small time interval dt, we can write the equation

$$dN = -\lambda N(t)dt \tag{28.1}$$

The minus sign takes into account the fact that the collection of radioactive nuclei is getting smaller over time.

By rearranging terms we can write this equation as

$$dN/N(t) = -\lambda dt \tag{28.2}$$

and proceed to take the integral of both sides of the equation so that we get

$$\int_{N(0)}^{N(t)} \frac{dN}{N} = -\int_{o}^{t} \lambda dt \tag{28.3}$$

Since the integral of $1/N$ is just the natural logarithm of N and the integral of dt is just t, this equation can be rewritten as

$$\ln N \Big|_{N(0)}^{N(t)} = -\lambda t \Big|_{0}^{t} \qquad (28.4)$$

28.6.1. Activity: The Mathematics of Radioactive Decay

a. Put in the limits of the integrals in Equation 28.4 and take the exponent of each side of the resulting equation to show that the number of radioactive nuclei remaining after a time t is given by the expression $N(t) = N(0)e^{-\lambda t}$ where $N(0)$ is the number of radioactive nuclei at $t = 0$.

b. How is the half-life, $t_{1/2}$, related to the decay constant, λ? You can derive this relationship easily by using Equation 28.4 and letting $N(t) = (1/2) N(0)$ when $t = t_{1/2}$. By using these procedures, show that $t_{1/2} = \ln 2/\lambda = 0.69/\lambda$.

c. Did you predict an inverse proportionality between $t_{1/2}$ and λ in Activity 28.5.2c?

d. Using the expression $N(t) = N(0)e^{-\lambda t}$ show that the counting rate dN/dt at time t is given by $dN/dt = -\lambda N(t)$.

28.7. IS CESIUM 137 REALLY LONG-LIVED?

Cesium 137 has a half-life of 30.2 years. You should be able to measure the counting rate from a sample of Cs 137 or some longer-lived source. Would you expect to be able to detect any change in counting rate during a one-hour period of counting? In order to complete this activity you will need:

- 1 computer data acquisition system
- 1 battery-operated radiation monitor
- Cs 137 source, 5 μ c (or a longer-lived source)
- 1 Scotch tape, 10 cm

Recommended group size:	4	Interactive demo OK?:	N

Before starting your measurements, place the radiation sensor at its usual location on the lab table and tape the radioactive source directly in front of the Geiger tube. Set the radiation software run time for one hour and the interval for one minute.

While the data are being recorded by the computer-based laboratory system you can read about radioactivity and its detection in the next few sections.

28.7.1. Activity: Counts from a Long-Lived Source

a. Do you predict that you will notice a decrease in counting rate during a one-hour period for your source? Explain.

b. Use the data acquisition software to record the number of counts/minute for 60 one-minute long trials (or use the experiment file L280701). At the end of the hour carefully transfer the data to a spreadsheet and calculate the average and standard deviation used in the spreadsheet. Summarize your average number of counts/minute and standard deviation in the space that follows.

c. In Unit 2 you explored the relationship between the standard deviation, σ, and the square root of the average number of counts in a counting interval. You should have found them to be about the same. Does this rule of thumb hold here? Compare the square root of the average number of counts/min to the standard deviation. Comment on the result of the comparison.

d. Study the spreadsheet. Is there any *statistically significant* decrease in the decay rate of the cesium during the hour over which you took measurements? Can you explain your observations on the basis of the mathematical theory given a collection of nuclei with such a long half-life? (By statistically significant we mean do the counts lie within one standard deviation of the average two-thirds of the time?)

28.8. THE ATOMIC NUCLEUS AND RADIOACTIVE EMISSIONS

To study the phenomenon of radioactivity, you need to know more about what physicists currently believe about the nature of atoms and their nuclei. Unfortunately, you lack the time and equipment to learn about these things by doing fundamental investigations. Thus, we're going to break from our usual pattern in this enterprise and tell you what's believed about the structure of atoms without answering the much more profound questions of why physicists believe what they believe about atoms.

All atomic nuclei are thought to consist of neutrons and protons held together, not by gravitational attraction or electric or magnetic forces, but rather by strong nuclear forces acting over a short range. The proton has the same magnitude of charge as the electron, but its charge is positive. It has a mass of almost 2000 times the mass of the electron. The neutron has a similar mass but no electronic charge. Protons in the nucleus ought to fly apart as a result of the Coulomb repulsion between them; the fact that they don't is a testimonial to the power of the even stronger nuclear forces that are *always* attractive and act on both protons and neutrons. The number of protons in the nucleus determines how many electrons are in the vicinity of an electrically neutral atom. Thus, a neutral atom has the same number of electrons surrounding the nucleus as it has protons in the nucleus. It is the number of electrons in a neutral atom that determines how the atom behaves chemically. Each chemical element is defined by the number of protons it has in its nucleus. For example, hydrogen has one proton, helium has two protons, lithium has three protons, and so on.

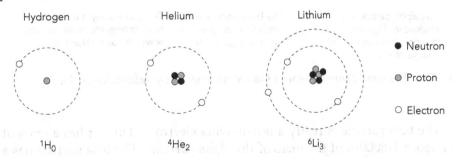

Fig. 28.4. The three lightest chemical elements showing the number of protons that define the element along with the most common number of neutrons in the nucleus. The gray circle represents a proton, the black circle a neutron, and the white circle an electron. The diagram is simplified since electrons are not pictured by physicists as orbiting the nucleus of an atom in nice, neat circles.

Although the chemical behavior of an element is keyed to the number of its electrons and hence also to the number of its protons, *different atoms of an element do not always have the same number of neutrons in the nucleus.* We call different nuclei types of an element *isotopes.* For example, even the very lightest element, hydrogen has three isotopes: hydrogen, deuterium, and tritium. This is shown in Figure 28.5.

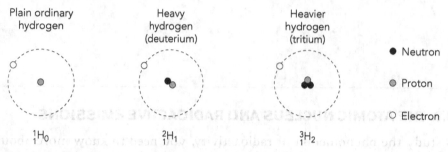

Fig. 28.5. Three isotopes of hydrogen. Deuterium is of great interest as a fuel for the nuclear fusion process, which may allow mankind to have an abundant source of low-cost energy.

Ordinary hydrogen only has one proton in its nucleus. Deuterium has one proton and one neutron. Tritium has one proton and two neutrons.

Certain isotopes are unstable and decay by spitting out energetic particles. These isotopes are said to be *radioactive.* When this decay occurs some of the neutrons or protons rearrange themselves or are transformed in some way. The three most common particles given off in the decay process are known as the alpha particle, the beta particle, and the gamma ray respectively. The alpha particle is not really a single particle. It consists of a collection of two neutrons and two protons. In fact, it is the nucleus of the element helium, which ordinarily contains two neutrons and two protons. Thus, it has two fundamental units of charge.

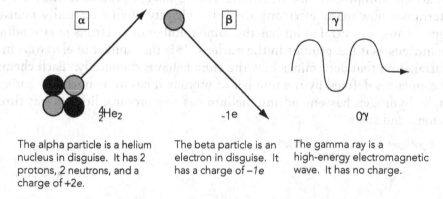

The alpha particle is a helium nucleus in disguise. It has 2 protons, 2 neutrons, and a charge of +2e.

The beta particle is an electron in disguise. It has a charge of −1e

The gamma ray is a high-energy electromagnetic wave. It has no charge.

Fig. 28.6. The most common types of radiation emitted by radioactive nuclei.

The beta particle is really a fast-moving electron. Thus, it has a mass of only about 1/8000th of the mass of the alpha particle. The beta particle has a negative charge of one unit. The gamma particle is a high-energy electromagnetic wave that consists of mutually perpendicular electric and magnetic fields that oscillate as it travels. It has momentum, but no mass or electric charge.

28.9. DETECTING IONIZING RADIATION WITH A GEIGER TUBE

The fact that alpha, beta, and gamma particles can ionize other atoms in their paths allows us to detect their passage electronically with a device called a Geiger tube. In a Geiger tube, an ionizing particle is passed into a cylinder of gas that has a high voltage between a central electrode and the outside of the cylinder. When electrons are knocked off atoms, the electrons flow toward the central electrode while the ions flow toward the outer wall of the cylinder. This creates a burst of current that can be amplified and sent to a computer or electronic counter so the passage of a particle of ionizing radiation can be recorded.

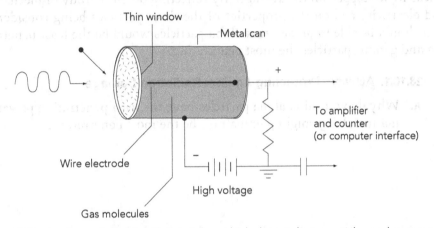

Fig. 28.7. A schematic for the Geiger tube, which detects beta particles and gamma rays. The radiation sensor has a Geiger tube and amplifier circuit in it.

A Geiger tube is best for detecting beta particles. Alpha particles are stopped in the thin window of mica at the end of the tube. Gamma rays travel very far without colliding with electrons and tend to pass on through the tube while beta particles ionize the gas in the tube.

28.10. THE HEALTH EFFECTS OF RADIATION

This is a vast and controversial subject about which literally thousands of books have been written. It is widely accepted that alpha, beta, and gamma particles that are emitted from radioactive nuclei can penetrate into the body and ionize atoms and molecules in the body. This changes the chemical bonds inside of living cells. If enough of these changes occur inside a cell, they can lead to the death of the cell, inaccurate reproduction of the cell, or malfunctioning of the cell. The degree of damage is believed to be roughly proportional to the amount of a given type of radiation a person is exposed to. It must be noted, however, that some scientists believe that very low amounts of radiation exposure will not cause permanent damage. They cite evidence that the body can repair itself. Others believe that any amount of radiation exposure can cause permanent damage. Currently, this remains an unsettled scientific controversy.

Based on studies of humans exposed to large amounts radiation and to animal studies in the laboratory, it is currently believed that radiation can increase the probability of an individual getting cancer five to twenty years after exposure and that radiation can cause mutations in the egg and sperm cells of potential parents that could lead to a higher incidence of genetic diseases in the population.

Ionizing radiation loses energy as it collides with atoms and molecules in its surroundings. It also fans out in all directions from the collection of radioactive nuclei that generate it. Shielding experiments indicate that typical alpha, beta, and gamma particles having approximately the same energy have quite different penetration power in materials. For example, alpha particles can be stopped by a sheet of paper or a couple of centimeters of air. Beta particles can be stopped by a thin book or about ten or twenty centimeters of air, while gamma rays might require several room lengths of air or several lead bricks to be stopped on the average. By reflecting on your study of mechanics and electricity and on the properties of the three particles being considered, you should be able to predict that alpha particles would be the least penetrating and gamma particles the most penetrating.

28.10.1. Activity: Protecting Against Radiation Damage

a. Why do you think alpha particles have the least penetrating power in matter? Why might gamma rays be the most penetrating?

b. What is more dangerous: swallowing a source of alpha, beta, or gamma particles? Give reasons for your answer.

c. What is more dangerous: having a source of alpha, beta, or gamma particles on the table in front of you? Give reasons for your answer.

d. Suppose you have a high-activity source of radiation on the table in front of you. Suggest two methods for protecting yourself from the radiation.

28.11. THE NUCLEUS AND RADIOACTIVE DECAY SERIES

When an alpha or beta particle leaves the nucleus, charge and energy are carried away from the nucleus. Physicists continue to believe that these quantities are always conserved, even in nuclear decay. Thus, when the beta particle carries a unit of negative electron charge away, the nucleus that is left behind must have an additional unit of positive charge. This will be the case if the beta decay occurs because a neutron has been transformed into a proton. A nuclear beta decay process is shown symbolically in the diagram below. Try to guess the meaning of the superscripts and subscripts. These will be explained in more detail below.

$$\text{Neutron} \qquad\qquad \text{Proton} \qquad\qquad \text{Electron}$$

$$^{1}_{0}n \longrightarrow {}^{1}_{+1}p \;+\; {}^{0}_{1}\beta$$

Fig. 28.8. Symbolic representation of a neutron in the nucleus being transformed into a proton while spitting out a beta particle.

When a decay occurs, the nucleus rearranges itself so that it has less energy. The difference in energy of a nucleus before and after decay is carried away by the departing particle.

Because of the conservation of charge we can set up a scheme to do bookkeeping and figure out what element and which isotope of that element is left behind after a decay of a given type. Figure 28.9 outlines how we can display the one- or two-letter abbreviation for each chemical element and the number of protons and neutrons in a given isotope of that element.

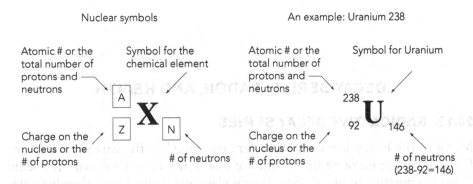

Fig. 28.9. Symbols to display the neutrons and protons in nuclear isotopes in general and for the uranium 238 in particular.

Let's use the bookkeeping method to display the isotope that results after an alpha decay of uranium 238. This is shown as follows.

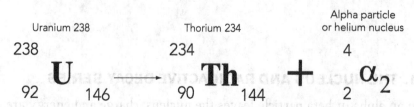

Fig. 28.10. Symbols to display the neutrons and protons when the uranium-238 nucleus gives off an alpha particle.

Since the alpha particle carries away two units of charge in the form of protons, the charge on the daughter nucleus must be 90. A look at the periodic table in any introductory physics or chemistry text reveals that thorium has 90 protons. Once the symbol for thorium, which is Th, is put in, the rest of the bookkeeping can be done easily. There are two fewer neutrons and the atomic number (that is, the number of protons and neutrons) is reduced by four units.

A similar job of bookkeeping can be done when thorium 234 undergoes a beta decay. This is shown in the diagram below.

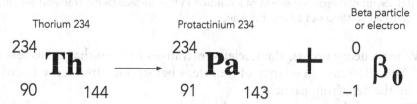

Fig. 28.11. Symbols to display the neutrons and protons when the thorium-234 nucleus gives off an beta particle.

DECAY SERIES, RADON, AND HEALTH

28.12. RADIOACTIVE DECAY SERIES

Many of the heavy elements that were contained in the earth when it was formed contain extra neutrons and are just dying to settle down into being lighter elements. Some of these heavy elements decay into a daughter element; the daughter in turn is the parent of another element in the decay chain and so on. For example, uranium 238 is the head of a radioactive decay chain in which a series of alpha and beta decays (followed by one or more gamma decays) lead ultimately to an isotope of lead that is stable rather than radioactive. The uranium decay chain is depicted on the next page. It is of special interest because one of the key daughters in the decay series, radon 222, is now recognized as having the widespread potential for increasing the lung cancer risk in people who live in certain unventilated basement areas.

28.12.1. Activity: Elements in the Radioactive Series

Study the following decay series. By using the conservation of charge for each type of radiation shown in Figures 28.6, 28.10, and 28.11, determine how many neutrons and protons the elements that go into each of the blank daughter rectangles should have. Check the periodic table and list of properties of elements contained in the appendix of an introductory physics textbook. Fill in the name, chemical element symbol, and atomic number for each of the "missing" elements on the chart that follows.

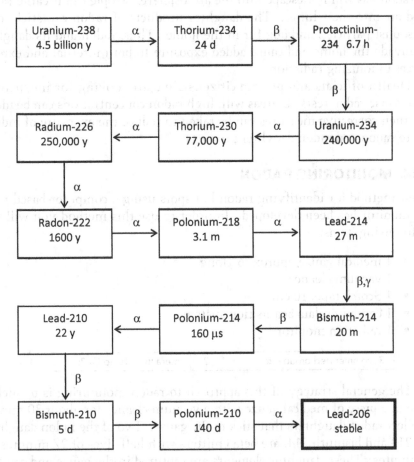

Fig. 28.12. The uranium–238 decay series shown above is one of several natural decay series that start with heavy isotopes of thorium or uranium.

28.13. THE RADON PROBLEM

Uranium 238 occurs naturally in soils. Some soils, like those in parts of Pennsylvania, contain more uranium than others. Uranium 238 has a half-life of over 10^9 years, a long time by our ordinary measures of time. It's going to stick around for a long, long time. As the uranium 238 steadily decays, thorium 234 is formed. The thorium 234 begets protactinium that begets uranium

234 that begets thorium 230 that begets radium 226 that, in turn begets radon 222, and so on. All the parents and grandparents of radon 222 stay in the soils. These elements bond chemically to soil particles. You are safe from these radioactive elements unless you make a practice of eating dirt.

Unlike all the other daughter elements in the uranium 238 decay series, radon is a chemically inert gas. The electrons surrounding the radon atom like to stick close to the radon nucleus. They do not form chemical bonds with other elements. The radon gas is free to seep up through the soil and into the atmosphere where it spreads out and becomes diluted. However, in enclosed basement areas, and especially in new, well-insulated houses, it is possible that the radon gas will not escape into the atmosphere. People can breathe radon-filled air into their lungs. The daughter products of radon can stick to the walls, dust particles, and dog fur in the home. Thus, radon and its daughters can invade the home and cause added exposure to both internal and external sources of ionizing radiation.

Health officials and private citizens alike are looking for inexpensive, easy-to-use radon tests. If areas with high radon concentrations can be identified, then corrective measures can be taken to reduce the exposure of individuals to radon and radon daughter products.

28.14. MONITORING RADON

A new method for identifying radon hot spots using a computer-based radiation monitor has been developed. In order to use this method you will need the following items:

- 1 medical gauze, approx. 6" long
- 1 vacuum cleaner
- 1 Scotch tape, 10 cm
- 1 computer data acquisition system
- 1 radiation monitor

Recommended group size:	2	Interactive demo OK?:	N

The general strategy of this approach to radon monitoring is to suck air through a piece of medical gauze with a vacuum cleaner for about 30 minutes to collect radon daughters that stick to the gauze. Two of the radon daughters, lead 214 and bismuth 214, are beta emitters with half-lives of 27 minutes and 20 minutes. These daughter elements are captured in the gauze and can then be transported to the radiation monitor immediately for counting. After recording counts from the gauze for about 4 hours, you should get data that yields a decay curve with a half-life somewhere in the 30-minute range.

Instructions for Activity 28.14.1:

1. *Reserve the equipment*: Sign up for an hour to use the vacuum cleaner and a basement area equipped with an electrical outlet.

2. *Practice Transferring Data from the Nuclear Counting Software to a Spreadsheet*: The computer-based radiation counting system uses software for the nuclear counting. If you haven't already been successful at transferring the counts/interval data obtained by the nuclear counting software to a spreadsheet, take some data and practice this maneuver. This is an essential skill without which you might lose all your data.

3. *Mount the Gauze on the Vacuum Cleaner*: Fold the gauze over several times and tape it to the intake end of the vacuum cleaner you have been assigned. Cover the intake with enough tape so that all the air coming into the vacuum cleaner must flow through the gauze. Leave at least a 1.5" × 1.5" area in which air can flow through the gauze so the vacuum cleaner doesn't overheat.

4. *Set Up Your Computer Data Acquisition System*:

   a. Set up an experiment file to count radiation (or use the file L281401), plug in the radiation monitor and interface. Make sure the system is counting properly.

   b. If needed, adjust the experiment file for 1 minute counting intervals and a run time of four hours. Place a DO NOT TOUCH sign on the computer with your names, the date, and the hours of use. You will need to reserve the computer for just under 5 hours.

5. *Collect the Radon Daughters*: Take the vacuum cleaner to the basement room you plan to monitor, plug it in and run air through the vacuum cleaner for exactly 30 minutes. Record the time of your data taking, any special conditions in the basement such as an open window, and its location as part of Activity 28.14.1.

6. *Race Back to the Computer-based Radiation Monitoring System*: As soon as the 30-minute collection period is up, run back to the lab with your sample. Fold the gauze up into a neat, small, flat wad. Tape the gauze directly against the Geiger tube.

7. *Collect Data on the Counting Rate vs. Time*: Collect 240 one-minute intervals over the next four hours. You can go away now and come back in four hours to save your data file. DON'T FORGET TO COME BACK. OTHERS MAY NEED THE COMPUTER!

8. *Transfer Your Data to a Spreadsheet*: Transfer your counts vs. seconds elapsed data into a spreadsheet and save the spreadsheet file. We will work with the data during the next activities.

28.14.1. Activity: Radon Monitoring Record

Record the following data:

Date of monitoring: _____

Location of basement (building, room number): _____

Special conditions (clean, dusty, ventilated, etc.)

Air sample: start time: _____   end time: _____

Counting: start time: _____   end time: _____

Number of counts/min in the first minute of counting: _____

Average background counts/min at location of counting: _____

Comments:

Note: It's very important for you to know how many minutes elapse between the time you stop collecting radon daughters by turning off the vacuum cleaner and the time you begin counting the radiation emitted from the gauze.

FITTING THE RADON DATA WITH A THEORY

28.15. DOES THE RADON DATA AGREE WITH THEORY?

The radon data you collected is not easy to explain theoretically right off the bat. Take another look at the uranium-238 decay series chart. Notice that the radon gas decays into four daughters before decaying to lead 210, which is essentially stable for our purposes. Each daughter has its own half-life and type of particle it gives off. Instead of having separate decay curves, we have linked decay curves. Thus, while the nuclei of a given daughter is decaying away as you count, the parent is creating new daughters of that type. What a headache! We are going to create the decay data for each of the daughters as if it were free to decay completely. We are then going to link all the simulated data to reflect the fact that each parent continues feeding its daughter. How closely will our theory of linked radioactive decay fit your experimental data? How significant is the radon level in the basement you monitored? Should people be living in that basement room or not?

How Much Radon Gas Did You Detect?

The activity of radon gas in air is measured in picocuries per liter where 1 picocurie (pCi) represents 2.22 disintegrations of radon per minute. The Environmental Protection Agency (EPA) has warned that people should not be living in or spending a significant amount of time in any space that has a radon level of 4.0 pCi/liter or more. The average home has a radon level of 1.4 pCi/liter. In roughly a million homes in the United States radon levels are 5 to 10 times higher than the EPA limit.

There are a number of techniques commonly used to determine radon levels in picocuries/liter. In order to have some estimate of what kind of radon level is in the space that you took your sample in, you will need to find a conversion factor between the counting rate you observed from the gooey gauze monitoring and the radon level in picocuries/liter. For example, suppose your instructor or someone in your class used another technique to find that a certain space had a radon level of 8.5 pCi/l. When using the gauze method in that same space you found that 10 minutes after the vacuum cleaner was turned off, the corrected counting rate, dN/dt, detected in the Geiger tube was 50 counts/minute. Then your conversion factor from counts/minute to pCi/l would be given by the ratio of the two factors as shown below.

$$F = [8.5 \text{ pCi/liter}]/[50 \text{ cts/min}] = 0.17 \text{ [pCi/liter]/[cts/min]}$$

28.15.1. Activity: Calculating the Radon Level

a. First let's determine the conversion factor, F, for your specific monitoring apparatus. Let C represent the number of pCi/liter you or your instructor obtained using a commercial monitoring device at a known location. Let G be the number of counts/minute detected by a radiation monitor during the eleventh minute (at 10 minutes) after the vacuum cleaner was turned off. Note both C and G must be determined by monitoring in the same space. Fill in the blanks below, if you can:

$$C = \text{_____} \text{ pCi/liter}$$

$$G = \text{_____} \text{ counts/minute}$$

$$F = C/G = \text{_____} \text{ [pCi/liter]/[cts/min]}$$

b. Open up your radon data spreadsheet and determine the average counting rate dN/dt in counts/minute during the tenth, eleventh, and twelfth minute *after the vacuum cleaner was turned off* for the particular space you were monitoring. Next, use this value to calculate the approximate radon level in picocuries per liter.

Approximately radon level =

$$F \times \frac{dN}{dt} = \text{\_\_\_\_\_} \times \text{\_\_\_\_\_} = \text{\_\_\_\_\_} \text{ pCi/liter}$$

c. With regard to its possible radon level and EPA warnings, would you be willing to live in the space you were monitoring?

d. Your results are very approximate! List some of the factors that might cause your result to fluctuate or be uncertain.

Graphing the Data for *dn/dt* vs. *t*

You will need to prepare your spreadsheet data so it can be graphed and compared with the theory that you will be developing.

28.15.2. Activity: Radon Monitoring Record

a. Open up your radon data spreadsheet. Create a time column (in minutes rather than seconds), subtract the average background cts/min (in the room where you are doing the counting with the radiation sensor) from the recorded counts/min, and put it in a column labeled corrected cts/min. **Note:** When making corrections, the number of counts/minute will be bouncing around so much due to statistical fluctuations that you may encounter zero or a negative number of counts in a given interval when background is subtracted. If this happens, simply set the corrected counts/minute to one. Why one? Well, one is pretty darn close to zero, and given the way things bounce around it's not a very big fudge. It's a fudge that will allow you to the take the natural log of one to get zero rather than the natural log of zero which will give you an impossible result like minus infinity (or $-\infty$ for short).

    b. Graph the data for the corrected value of dN/dt in counts/minute as a function of minutes. Affix the graph in the space that follows.

28.16. DEVELOPING A MODEL TO EXPLAIN YOUR DECAY CURVE

Typically, in a poorly ventilated basement room, the radon gas that diffuses up from the ground floats around in the room. Within 3.8 days approximately half the radon 222 atoms that come into the room at about the same time give off alpha particles and decay into polonium 218. Half of the polonium 218 decays into lead 214 within 3.1 minutes of the time it is produced, and so on. After a few hours the room contains a mixture of radon 222 and its daughters. If you were to develop a spreadsheet model of this process, you would find that after a few hours the radon would reach *secular equilibrium* with its daughters Po 218, Pb 214, Bi 214, and Po 214. A state of *secular equilibrium is defined as the state in which each of the daughter products has the same activity or decay rate as the parent nucleus.* For the purposes of developing a model to describe our radon monitoring data, we only care about what happens over time to the two daughters that give off beta particles Pb 214 and Bi 214.

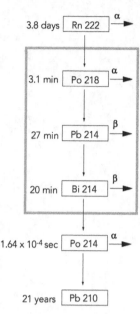

Fig. 28.13. The portion of the uranium 238 decay series of interest in radon monitoring.

What's the relative number of radon nuclei compared to the number of daughters within a few hours after the radon enters the room? In this scenario there are originally only radon atoms in the room.

28.16.1. Activity: Relative Activity of Radon Daughters

a. Review the results you obtained in Activity 28.6.1. What is the relationship between the decay constant, λ, and the half-life, $T_{1/2}$, of a radioactive isotope? Calculate the decay constant in terms of the probability of decay per minute for the radon nucleus and all the daughter nuclei shown in Figure 28.13.

$$\lambda_{Rn222} = \underline{\hspace{2cm}} \quad / \min$$

$$\lambda_{Po218} = \underline{\hspace{2cm}} \quad / \min$$

$$\lambda_{Pb214} = \underline{\hspace{2cm}} \quad / \min$$

$$\lambda_{Pb210} = \underline{\hspace{2cm}} \quad / \min$$

b. Review the results you obtained in Activity 28.6.1. What is the equation for the activity or decay rate, dN/dt, as a function of the decay constant, λ, and the total number of radioactive nuclei, $N(t)$, present at time t?

c. Suppose there are 10 million radon atoms present in a liter of air in a room. Calculate the activity (or decay rate) for the radon 222.

d. If the radon is in secular equilibrium with its daughters, what are the numbers of each of the daughter atoms? **Hint:** For Po 218 you should get $N = 5754$. What are the values of N for Pb 214 and Bi 214?

One more thing before getting on to the development of our spreadsheet model. Let's consider more carefully what you would expect to happen as particles of dust in a basement room accumulate on the gooey gauze while you run the vacuum cleaner.

28.16.2. Activity: What's in the Dust?

Do you expect that radon atoms stick to particles of dust in the room? How about the polonium, lead, or bismuth daughters? Will they stick to dust particles in the room? Explain.

Okay, let's assume you've flipped off the vacuum cleaner and rushed over to start counting your sample immediately. Assuming no radons have stuck to the gauze, then we have a mixture of just the daughter products. Assume that you have 5754 polonium 218's in your piece of gauze and the appropriate number of other daughter atoms as calculated in Activity 28.13d. If the daughter elements are in secular equilibrium, let's figure out how one element now decays into another as time goes by and how many beta particles will be emitted each minute.

28.16.3. Activity: The Radon Decay Model

a. Suppose at time $t = 0$ there are 5754 polonium 218 atoms present in a liter of air in a room and 48688 atoms of lead 214. What is the activity or decay rate at time $t = 0$ min for the polonium 218? How many polonium nuclei are left at the end of the first minute? How many lead 214 nuclei exist at the end of the first minute because they have been produced in that minute?

b. Let's check up on the lead 214 at the end of the second minute. How many lead 214 nuclei are lost due to decay? How many lead 214 nuclei are gained because of the decay of its parent atom polonium 218?

c. If it's available, open a pre-prepared spreadsheet entitled "S281603.XLS." Study the first few rows. How do the numbers you just calculated in parts a. and b. compare with the ones in the spreadsheet for the number of each element?

d. Hey, this spreadsheet is supposed to show a decay curve for the two daughters that emit beta particles! How come the counting rate for total betas is going up at first instead of down? **Hint:** What is happening to the polonium 218 at a rapid rate at first? How does this influence the lead 214 decay rate? Did you notice any evidence of this predicted upswing in the plot of your data requested in Activity 28.14.1b?

e. Mess around with the spreadsheet model and change the original number of polonium atoms until the total beta decay counting rate at time $t = 0$ (from the time you turned off the vacuum cleaner) matches the value you obtained. Then graph the natural log of dN/dt vs. time. Label this as a model decay curve or a theoretical curve. Next do a similar plot of your data using exactly the same scale. Insert the two graphs behind this page. How do they compare? **Note:** If you monitored a low radon area and don't have a decent decay curve to look at, you should analyze data provided by the instructor.

INDEX